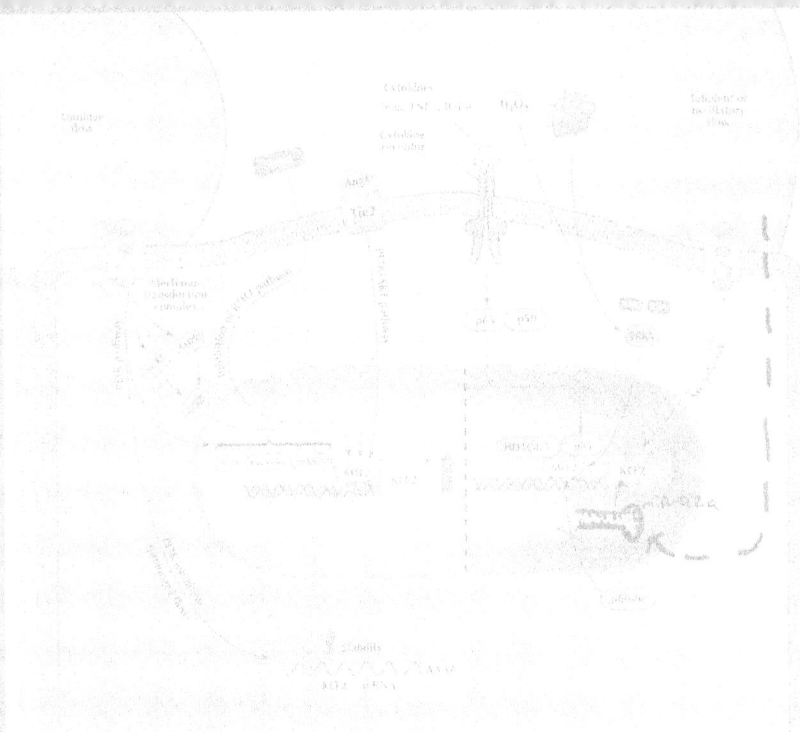

MODERN MECHANOBIOLOGY

MODERN MECHANOBIOLOGY

Convergence of Biomechanics, Development, and Genomics

edited by
Juhyun Lee | Sharon Gerecht
Hanjoong Jo | Tzung Hsiai

JENNY STANFORD
PUBLISHING

Published by

Jenny Stanford Publishing Pte. Ltd.
Level 34, Centennial Tower
3 Temasek Avenue
Singapore 039190

Email: editorial@jennystanford.com
Web: www.jennystanford.com

British Library Cataloguing-in-Publication Data
A catalogue record for this book is available from the British Library.

Modern Mechanobiology: Convergence of Biomechanics, Development, and Genomics

ISBN 978-981-4800-58-7 (Hardcover)
ISBN 978-0-429-29483-9 (eBook)

Contents

Preface

Mechanobiology has been studied and applied in various research disciplines in recent decades. This is because of providing a solution of different or combination signal transductions mechanically with chemical reactions in molecular events. Therefore, simple to advanced engineering techniques have been introduced to investigate mechanosignal transduction underlying tissue development, injury, and recovery in order to understand disease models. As it becomes a field of convergence of science, this book covers basic concepts from mechanotransduction to modern advanced research directly related to the disease model using different technologies.

This book represents an attempt to cover recent mechanotransduction research and to introduce the direction of the current research trend to biomedical researchers. Researchers from many different institutions have contributed to individual chapters of the book, enriching the interface between mechanobiology and precision medicine for personalized diagnosis and intervention. The first two chapters form the fundamental basis of vascular disease in response to hemodynamic shear stress. Subsequent chapters embark on a journey of cardiovascular development and regeneration, valvular and cardiac morphogenesis, mechanosensitive microRNA and histone deployment, computational fluid dynamics, and light-sheet imaging. The book represents a paradigm shift from traditional biomechanics and signal transduction to transgenic models, including new zebrafish and chicken embryos. This edition targets a wider readership from academia to industry and government agencies in the field of mechanobiology.

We hope this book will help students and biomedical researchers understand the fascinating world of mechanotransduction and develop and pursue the next advanced topic in detail. We express our utmost gratitude to the authors for agreeing to the publication of this book.

Juhyun Lee

2021

Chapter 1

Shear Stress, Mechanosensors, and Atherosclerosis

Suowen Xu and Zheng Gen Jin
Aab Cardiovascular Research Institute, Department of Medicine, University of Rochester, 601 Elmwood Avenue, Rochester, NY, 14623, USA
Suowen_Xu@URMC.Rochester.edu; Zheng-Gen_Jin@URMC.Rochester.edu

Endothelial cells line the innermost layer of a blood vessel and regulate multiple functional aspects of vascular homeostasis, such as vascular tone, inflammation, vessel permeability, and angiogenesis. Endothelial cells possess multiple specialized mechanosensing molecules or microstructures (termed "mechanosensors") that sense different patterns of blood flow (in a concerted manner). Mechanosensing is a prerequisite step for integrating the mechanosignal into endothelial cells and leads to different patterns of endothelial gene expression. Emerging evidence using genetically engineered mice and pharmacological modulators has shown that these mechanosensors are critical regulators of endothelial function and probably the pathogenesis of atherosclerosis-related cardiovascular diseases. In this chapter, we delineate the specific

Modern Mechanobiology: Convergence of Biomechanics, Development, and Genomics
Edited by Juhyun Lee, Sharon Gerecht, Hanjoong Jo, and Tzung Hsiai

ISBN 978-981-4800-58-7 (Hardcover), 978-0-429-29483-9 (eBook)
www.jennystanford.com

role of each identified mechanosensor in vascular endothelial cell biology and atherosclerosis and highlight the possibility of targeting these mechanosensors to treat atherosclerosis.

1.1 Introduction

Cardiovascular diseases caused by atherosclerosis represent the main cause of global morbidity and mortality [1, 2]. Atherosclerosis is a chronic inflammatory disease [3] and a focal disease in which atherosclerotic plaques preferentially developed in regions of disturbed/turbulent flow (such as arterial branching points, bifurcations, and inner curvature) but less in regions of laminar flow (such as thoracic aorta) [4–8]. Therefore, the development of atherosclerosis itself is heterogeneous, and this focal nature of atherosclerosis makes the study of different patterns of hemodynamic forces an effective therapeutic strategy to combat atherosclerosis.

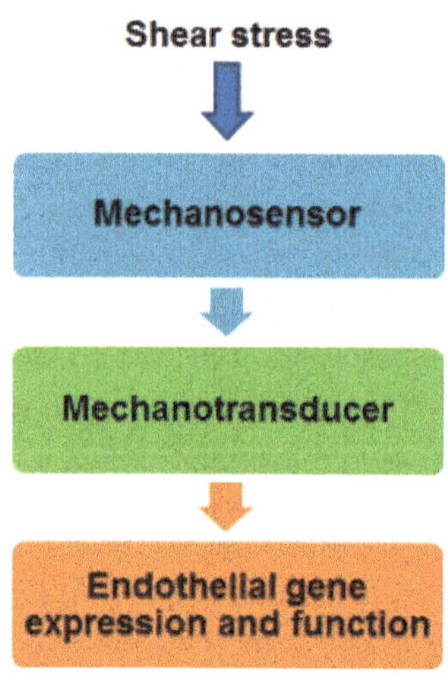

Figure 1.1 Mechanical regulation of endothelial gene expression. After exposure to different types of blood flow, the mechanical signal was sensed by mechanosensors existing on the endothelial cell membrane or specific microdomains. Then, the signal was further transduced to mechanotransducers, which relay the mechanosignal to the inside of endothelial cells. The eventual outcome is the activation or inhibition of gene expression program depending on specific transcriptional factors. By doing so, mechanosensors regulate important endothelial functions and the development of cardiovascular diseases.

Endothelial cells (ECs) are one major type of vascular cells. They line the intima by forming an endothelial monolayer, thereby separating circulating blood from human body [9]. In addition to being a natural barrier between blood components and vessel wall, they are the specific cell type that are in direct contact with blood flow and vessel wall. On the endothelial surface, there are many sensing molecules or microdomains that can serve as effective mechanosensors. They sense and transduce the mechanosignal exerted by shear stress to the inside of ECs. Dysfunctionality of these sensors leads to the alteration of endothelial pathways, gene expression, and endothelial functions, such as vascular tone, vessel stability, and leukocyte adhesion (Fig. 1.1). In this chapter, we provide a systematic overview of the up-to-date functions and mechanisms of different mechanosensors in mediating endothelial function and atherosclerosis. We also highlight future direction in mechanosensor-based studies, without an aim to accelerate cardiovascular drug discovery.

1.2 Shear Stress and Endothelial Phenotype

Generally, there are two major types of shear stress: one is the laminar flow and the other is disturbed flow. The different flow patterns were generated by different hemodynamic forces using different shear stress systems, such as plate viscometer, flow chamber, ibidi system, and other microfluidic systems. Mounting evidence in the past has shown that laminar flow protects against endothelial dysfunction and atherosclerosis, by promoting an endothelial protective and atheroprotective transcriptional program, mediated by transcriptional factors, such as Krüppel-like factor-2 (KLF2) [10–12], KLF4 [13–15], and Nrf2[1] [16]. In contrast, disturbed flow promotes endothelial dysfunction and atherosclerosis via atherogenic transcriptional factors, such as nuclear factor kappa light chain enhancer of activated B cells (NF-κB) [17], yes-associated protein (YAP) [18–20], and HIF1[2] [21, 22] (Fig. 1.2). Many drugs, such

[1]Nuclear factor erythroid 2–related factor 2.

[2]Hypoxia-inducible factor-1.

as statins [23, 24], resveratrol [24–26], tannic acids [27], and histone deacetylase inhibitor SAHA[3] [28], mimic laminar flow response by upregulating KLF2 expression and activity. On the other hand, many proatherogenic factors, such as lipopolysaccharide (LPS) [29] and tumor necrosis factor-alpha (TNF-α) [30], mimic the disturbed flow–induced response by activating NF-κB and downregulating KLF2 expression, suggesting the possibility of target mechanosensitive transcriptional factors modulating atherosclerosis.

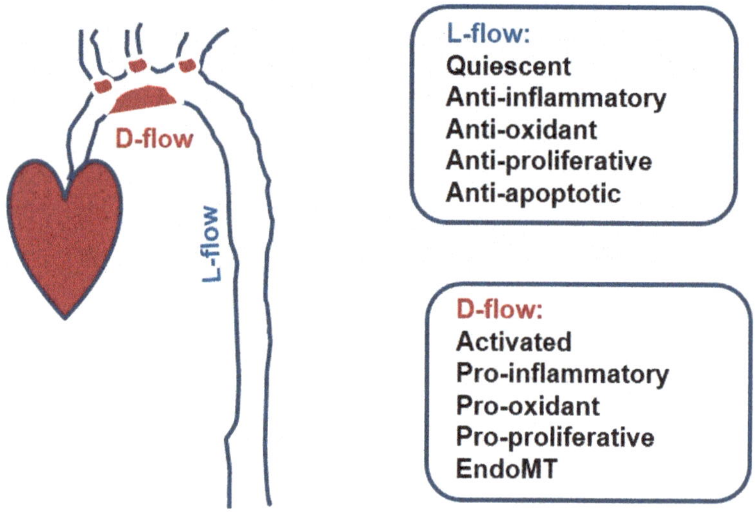

Figure 1.2 Focal development of atherosclerotic plaques in disturbed flow area, rather than the laminar flow area of the aorta. Two major hemodynamic forces exist in the vasculature: atheroprotective laminar flow and atheroprone disturbed flow. Two different types of blood flow elicit different gene expression patterns and transcriptional programs. Laminar flow leads to an antioxidant, anti-inflammatory, antiproliferative, and antiapoptotic program. The net effect of laminar flow is EC quiescence. In contrast, disturbed flow confers a proinflammatory, pro-oxidant, proproliferative response, and enhanced endothelial mesenchymal transition (EndoMT) in ECs, with the net effect of EC activation. Mitigating disturbed flow–mediated proatherogenic events or mimicking laminar flow–induced atheroprotective signaling could lead to the identification of novel therapeutic targets and strategies of atherosclerosis. Images were manually drawn using Microsoft PowerPoint.

[3]Suberoyl+anilide+hydroxamic acid.

1.3 Mechanosensors in Atherosclerosis

To date, there are several mechanosensors identified in ECs [31–34] that are responsible for mechanosensing and mechanotransduction (Fig. 1.3). We will discuss them in detail and provide an updated review next.

Figure 1.3 Known mechanosensors in endothelial cells. Current identified mechanosensors include the PECAM1/VEGFRs/VE-cadherin mechanosensing complex; ion channel TRPV4 and Peizo1; Rap1; integrins; GPCR; and some microdomain structures in endothelial cells, such as caveolae, primary cilia, and glycocalyx. These mechanosensors have an important pathophysiological role in regulating EC function and atherosclerosis development.

1.3.1 PECAM1/VEGFR2/VE-Cadherin Mechanosensing Complex

Platelet EC adhesion molecule-1 (PECAM1), vascular endothelial (VE) growth factor receptors (VEGFRs), and VE-cadherin are important proteins localized at EC–EC junctions. They serve as the primary mechanosensors in ECs. PECAM1 was rapidly phosphorylated (at tyrosine residues) after exposure to shear stress [35]. Then, the phosphorylation of Akt and endothelial nitric oxide synthase (eNOS) was increased [35]. Depletion of PECAM1 almost abolished flow- and exercise-induced Gab1[4] phosphorylation at Y627, indicating the essential role of PECAM1 in mechanosensing [36].

Similar to PECAM1 tyrosine phosphorylation, flow also rapidly induces VEGFR2 phosphorylation [37]. In 2008, Tzima et al. discovered a mechanosensitive complex composed of PECAM1/VEGFR2/VE-cadherin that is responsible for mechanosensing and mechanotransduction [38]. Classically, PECAM-1 induced Src kinase–dependent phosphorylation and transactivation of VEGFRs. VE-cadherin could possibly function as an adaptor [37]. Recent studies have shown that flow shear stress induced VE-cadherin phosphorylation at Tyr 658 [37]. Recently studies have also revealed that VEGFR3 also forms a complex, thus adding a new component in the mechanosensing complex [39, 40]. In terms of the role of PECAM1 in atherosclerosis, in 2008, two independent groups independently showed in two different mouse models of atherosclerosis (apolipoprotein E knockout [$ApoE^{-/-}$] mice and low-density-lipoprotein receptor–deficient [$LDLr^{-/-}$] mice), that PECAM1 affects atherosclerosis in a site-dependent manner, suggesting that PECAM1 could perform dual functions depending on the patterns of blood flow [41, 42]. Specifically, PECAM-1 deletion inhibits NF-κB-dependent VCAM-1[5] expression when exposed to disturbed flow [41].

1.3.2 TRPV4

Transient receptor potential vanilloid subtype 4 (TRPV4) is a nonselective cation channel that regulates multiple facets of endothelial function (i.e., calcium influx, eNOS phosphorylation, and nitric oxide [NO] production) [43]. A recent study by Ye et al. has shown that TRPV4 is involved in irisin-induced endothelium-dependent vasorelaxation by enhancing extracellular Ca^{2+} influx in rat mesenteric arteries [44]. He et al. revealed decreased interaction between TRPV4 and KCa2.3 (Ca^{2+}-activated potassium

[4]GRB2-associated binding protein 1.

[5]Vascular cell adhesion molecule-1.

channel 3) in ECs from hypertensive mice and patients. The authors also discovered a new small-molecule compound, JNC-440, which potentiated TRPV4/KCa2.3 interaction in ECs, promoted vasodilation, and lowered blood pressure in mice without systemic activation of both channels [45]. Caires et al. recently showed that metabolism of omega-3 fatty acids is essential for TRPV4 activity and omega-3 PUFAs[6] increased TRPV4 channel activity in human ECs without affecting TRPV4 expression or trafficking [46]. TRPV4-dependent Ca^{2+} influx also mediates H_2S-induced vasodilation in ECs [47]. Moreover, shear stress sensitizes TRPV4 agonist–induced response by promoting the trafficking of intracellular TRPV4 to the plasma membrane, through exocytosis. In addition, shear stress enhances tyrosine phosphorylation of TRPV4 at site Y110 [48].

Recent studies have shown that TRPV4 was downregulated by oxidized low-density lipoprotein (oxLDL), but not native low-density lipoprotein (LDL) [49], thus regulating foam cell formation. Another study has also shown the genetic deletion or pharmacologic inhibition of TRPV4-blocked oxLDL-induced lipid uptake and foam cell formation in macrophages, independent of CD36, one major scavenger receptor responsible for lipid uptake [50]. These findings suggest that differential functions of TRPV4 may exist in regulating atherosclerosis. Further studies using tissue-specific knockout mice will give more direct evidence of the specific roles of endothelial or macrophage TRPV4 in atherosclerosis. A recent study has shown that pharmacological activation of TRPV4 by GSK1016790A (by oral gavage) inhibits atherosclerosis in *ApoE*$^{-/-}$ mice by promoting eNOS phosphorylation (at Ser1177) and activity. GSK1016790A also inhibits leukocyte adhesion to activated endothelium in vitro and in vivo [51]. It remains to be studied whether endothelium-specific overexpression of TRPV4 will also contribute to improved atheroprotection.

1.3.3 Piezo1

Piezo1 (also known as Fam38a) protein is a newly identified mechanosensing subunit of calcium-permeable nonselective ion channel expressed in ECs [52]. Mutations in Piezo were linked

[6]Polyunsaturated fatty acid.

to hereditary xerocytosis and muscular contracture-related syndromes [53]. A recent study by Zhang et al. has identified that the sarcoplasmic/endoplasmic reticulum Ca^{2+} ATPase interacts with Piezo and suppresses Piezo activity by acting on its linker domain, thus providing mechanistic insights into the regulation of the Piezo channel [54]. Total or endothelial-restricted knockout of Piezo1 was embryonically lethal in mice [55, 56]. Piezo1-depleted or Piezo1-deficient ECs show reduced calcium influx into ECs and VEGF-induced eNOS phosphorylation at Ser1177, reduced angiogenesis, and impaired alignment to the direction of laminar shear stress [55, 56]. In addition, a recent study has shown that Yoda1, which is a selective Piezo1 activator [57] that triggers calcium influx and NO as well as ATP production, mimics fluid flow to increase the phosphorylation of flow kinases Akt and ERK1/2 in human ECs in a Piezo1-dependent [58] and Piezo1-independent manner [59]. The exact therapeutic effects of Yoda1 on endothelial function and murine atherosclerosis remain to be further examined in mouse models of atherosclerosis. Moreover, the expression pattern and function of Peizo in macrophage and smooth muscle cells remain largely unknown. Due to the lethality of total or endothelial-specific deletion of *Piezo1* in mice, no data are available as to the role of Piezo in atherosclerosis in experimental animal models or human patients. The creation of inducible knockout of *Piezo1* in hyperlipidemic *ApoE*$^{-/-}$ or *LDLr*$^{-/-}$ mice will be important to investigate the role of Piezo in atherosclerotic cardiovascular diseases.

1.3.4 Primary Cilia

Primary cilia are 3–5 μm long sensory microtubule-like microstructures that protrude into the vessel lumen and respond to biochemical, biophysical, and biomechanical stimuli [33]. Primary cilia have long been considered as a mechanosensing structure that serve as mechanosensors [60] that are critical for calcium influx and the production of NO [61]. Primary cilia were also regarded as a magnifier of mechanosignal [33, 62]. It has been shown that nonciliated cells or cells undergoing chemical removal of cilia show a decrease in shear-responsive transcriptional factor KLF2 [63]. Recent studies have shown that the presence of primary cilia protects against the development of atherosclerotic lesions, evidenced by the deletion of the ciliogenesis gene *Lft88*, promotes vascular inflammation, and inhibits the activity of eNOS [64]). ECs lacking primary cilia are more susceptible to calcification [65]. Primary cilia are distributed in mouse aorta in an non-uniform manner, with preferential distribution in regions of disturbed flow (predilection sites) [62] but less in atheroprotective regions (where the flow pattern is laminar flow) of mouse aorta. On the basis of the atheroprotective profile of primary cilia, it would be interesting to develop new drugs that can increase the cilia number or activity.

1.3.5 Caveolae

Caveolae are special sphingolipids and cholesterol-rich microdomains that have an important role in mechanotransduction and blood vessel function [66]. Caveolae are enriched in mechanosensitive cell types, such as ECs. Caveolin-1 (Cav-1), Cav-2, and Cav-3 are three major coat protein components of caveolae [66]. An earlier study has shown that the density of endothelial caveolae in the vessel lumen was increased in response to flow [67]. Cav-1 is essential for flow-induced vasodilation, as Cav-1-deficient mice showed reduced eNOS-dependent vasodilation in carotid artery [68], which was rescued by endothelium-specific overexpression of Cav-1 [69], Interestingly, Cav-1 deficiency reduces atherosclerosis development [70], while endothelium-specific overexpression of Cav-1 increases atherosclerosis in $ApoE^{-/-}$ mice [71] by affecting vascular inflammation, NO production, and the influx of leukocytes. Further studies are needed to verify the function of other coat proteins in mechanosensing and mechanotransduction.

1.3.6 Rap1

Rap1 is recognized as a structural component of a telomere [72]. Recent studies from the Chrzanowska-Wodnicka group have elegantly shown that Rap1, a small GTPase protein, functions as a novel mechanosensor in ECs [73, 74]. The importance of Rap1 was shown by the decreased NO production in Rap1-deficient ECs in response to flow shear stress [73, 74]. Keeping with these data, the deletion of Rap1 impaired acetylcholine-induced, endothelium-dependent relaxations due to activation of NADPH oxidase–induced reactive oxygen species generation and reduced redox enzymes, thioredoxin 1, and glutaredoxin 1 [75]. In addition, Rap1 activation inhibits vascular hyperpermeability induced by VEGF or TNF-α [76]. These literatures indicate that Rap1 plays an important role in regulating vascular homeostasis. In contrast, a recent study has shown that Rap1 is elevated in human atherosclerotic plaques and regulates the production of proinflammatory cytokines in LPS-stimulated macrophages by promoting NF-κB activation, and this regulation is only observed in M1 macrophages (proinflammatory), but not M2 macrophages (anti-inflammatory), nor ECs and smooth muscle cells, suggesting the differential role of Rap1 in macrophages and other vascular cells [77].

1.3.7 Glycocalyx

Glycocalyx is an important microstructure existing in the apical surface of ECs. This microstructure serves as a barrier as well as a critical regulator in vascular development, mechanotransduction, vascular permeability, coagulation and leukocyte adhesion, and vascular homeostasis [78, 79]. It is considered as the first line of defense and consists of multiple negatively charged glycosaminoglycans and proteoglycans, including chondroitin sulfate, syndecans, heparan sulfate, and glypican-1 [80–83]. The disruption of glycocalyx structure results in increased vascular inflammation, edema, and vascular permeability [84]. It has been recently demonstrated that glycocalyx components, such as glypican-1, heparan sulfate, syndecan-1, and syndecan-4, are critical for flow-induced EC alignment to shear flow direction and NO production [80, 81]. Moreover, key glycocalyx protein, heparan sulfate proteoglycans, redistribute after shear, suggesting that the presence of glycocalyx could serve as a cell-adaptive mechanism in response to shear [83]. In contrast, the removal of the glycocalyx by heparinase III significantly impaired ECs to align to flow direction [83].

1.3.8 Integrins

Integrins are also specialized mechanosensors on the EC membrane surface that connect extracellular matrix and cytoskeleton [85,

86]. They are formed as a heterodimer of α and β subunits, which is critical for ligand binding, EC/extracellular matrix interaction, and intracellular signaling [33]. ECs plated on different cellular matrixes (such as collagen, fibronectin, fibrinogen, laminin, and gelatin) respond differently in response to various patterns of shear stress in terms of regulation of different integrin isoforms [33, 87, 88]. For example, a recent study by Yang et al. has shown that ECs plated on fibronectin and laminin respond differently to disturbed shear stress in patterns of the activation of integrins (β1 and β3), transforming growth factor-β/smad2, and NF-κB-dependent inflammatory gene expression [89]. Integrin β1 also serves as a membrane receptor for guidance molecule Semaphorin 7A and is involved in Semaphorin 7A–mediated proatherogenic events via the FAK[7]/MAPK[8]/NF-κB pathway [90]. Although the application of shear stress does not directly activate integrins, integrins do share the property of mechanosensors, possibly specialized mechanosensors [33]. In 2013, Yang et al. showed that shear stress increases apical eNOS activity via integrin β1 and caveolae [91]. A recent study has shown that fluid shear stress promotes the interaction between integrin β3 and Gα13, inhibiting RhoA and YAP activation [19]. Moreover, pharmacological activation of integrin by giving $MnCl_2$ to hyperlipidemic mice decreased atherosclerotic plaque formation [19], indicating the therapeutic potential to treat atherosclerosis by targeting integrin β3 and/or integrin β3/Gα13 interaction.

1.3.9 GPCR

G-protein-coupled receptors (GPCRs), such as S1P receptor, bradykinin B2 receptor, angiotensin II type I receptor, apelin receptor (APJ or APLNR), and Gαq/11, are also involved in mechanosensing and mechanotransduction alone or with other mechanosensors [31–33, 92]. Previous studies have shown that Gαq/11 forms a mechanosensitive complex with PECAM1, while, this complex is rapidly dissociated after flow induction [93]. Moreover, Gαq/11/PECAM1 interaction can be regulated by heparan sulfate, an important component in glycocalyx [94]. More

[7]Focal adhesion kinase.

[8]Mitogen-activated protein kinase.

recently, a similar phenomenon has been observed for APJ/PECAM1 interaction in regulating cytoskeletal remodeling adaptation to fluid flow [92]. In addition, the GPCR-independent effect of Gαq/11 on mechanosensing/mechanotransduction has also been recently observed in ECs [95]. The precise role of different GPCR proteins in mechanosensing and mechanotransduction warrants further evaluation in vitro and in vivo.

1.3.10 Emerging New Mechanosensors

In addition to the mechanosensors described above, there are some new mechanosensors emerging from translational research. For example, a recent study has shown that Notch1 is mechanosensitive and regulates junctional integrity and antiproliferation and anti-inflammatory cell recruitment [96]. Most importantly, endothelium-specific deletion of Notch1 aggravates diet-induced atherosclerotic lesion formation in mouse descending aorta, where the flow pattern is laminar flow. Therefore, Notch1 is a new mechanosensor in arterial endothelium, which is mechanosensitive and atheroprotective [96].

LOX-1 (lectin-like oxLDL receptor-1, also known as OLR1) is a major receptor of oxLDL in ECs [97], which is implicated in the initiation and progression of atherosclerosis [98]. A recent study has shown that disturbed flow (±5 dyne/cm^2) and high shear stress (120 dyne/cm^2) upregulate LOX-1 expression via AP-1, while laminar flow (20 dyne/cm^2) downregulates LOX-1 expression via a KLF2-dependent mechanism [99]. In terms of the fact that LOX-1 is membrane-localized protein in mechanotransduction, it is plausible that LOX-1 is a potential mechanosensor downstream of the mechanosensing complex. Gain- and loss-of-function studies in genetically engineered hyperlipidemic mice indicate that LOX-1 in WCs promotes the development of atherosclerosis.

The evidence summarized here indicates that effective mechanosensing and mechanotransduction require the concerted actions of known or even unknown mechanosensors. The discovery of new mechanosensors will further our understanding of endothelial mechanobiology and could possibly lead to new cardiovascular therapeutics.

1.4 Conclusions and Perspectives

Atherosclerosis is a complex disease involving multiple cell types and mechanisms, including chronic inflammation, oxidative stress, immunity disorder, and epigenetics [100–105]. The development of atherosclerosis is influenced by several factors, including diet, lifestyle, genetics, environment, and multiple risk factors [106, 107]. Among these risk factors, disturbed hemodynamic forces is an important one that triggers atherosclerosis. ECs are the major cell type in the vessel wall that sense mechanical force, including shear stress, which is exerted by the frictional force of the flowing blood. In the past decade, many flow-related mechanosensors have been identified as critical regulators of EC function and atherosclerosis [31–33]. Although much progress has been made in understanding the role of these mechanosensors in vascular biology, there are still several issues we need to dissect in future studies [31–33]. Some of these issues are:

- How these mechanosensors function in concert with sense and transduce the mechanosignal to the inside of the EC
- How they coordinate with mediate shear stress–mediated functional readouts, such as cell behavior change (proliferative or quiescent). The recent description of the caveolae/TRPV4/KCa2.3 channel complex in regulating shear stress–mediated vasodilation in coronary arteries is a good example [82].
- What the potential effects are of these mechanosensors in mediating the cross talk between ECs and circulation monocytes and the neighboring smooth muscle cells.
- How to identify new mechanosensors in ECs and characterize the biological functions of these new mechanosensors. The recent discovery of GPR68 as a flow sensor that mediates vasodilation and vascular pathophysiology provides a good example of utilizing high-throughput RNA library screening as a putative strategy [108].
- How to translate mechanosensor-related vascular biological findings to novel cardiovascular drug discovery.
- How to assess the pharmacological actions of a mechanosensor modulator in endothelial dysfunction and atherosclerosis.

Progress in this direction will allow us to develop pharmacological therapeutic agents targeting mechanosensors to treat atherosclerosis without doing exercise (i.e., mechanosensor modulators can be developed as exercise mimicking drugs or exercise pills). To address these questions and beyond, we need to be equipped with skills in cell biology, physiology, biomedical engineering, pharmacology, and functional genetics and genomics to explore the therapeutic potential of targeting mechanosensors.

References

1. Atlas Writing, G., Timmis, A., Townsend, N., et al. (2018). European society of cardiology: cardiovascular disease statistics 2017, *Eur Heart J*, **39**, pp. 508–579.
2. Benjamin, E. J., Virani, S. S., Callaway, C. W., et al. (2018). Heart disease and stroke statistics-2018 update: a report from the American Heart Association. *Circulation*, **137**, pp. e67–e492.
3. Hansson, G. K. (2017). Inflammation and atherosclerosis: the end of a controversy. *Circulation*, **136**, pp. 1875–1877.
4. Chiu, J. J. and Chien, S. (2011). Effects of disturbed flow on vascular endothelium: pathophysiological basis and clinical perspectives. *Physiol Rev*, **91**, pp. 327–387.
5. Rezvan, A., Ni, C. W., Alberts-Grill, N., et al. (2011). Animal, in vitro, and ex vivo models of flow-dependent atherosclerosis: role of oxidative stress. *Antioxid Redox Signal*, **15**, pp. 1433–1448.
6. Davies, P. F. (2009). Hemodynamic shear stress and the endothelium in cardiovascular pathophysiology. *Nat Clin Pract Cardiovasc Med*, **6**, pp. 16–26.
7. Abe, J. and Berk, B. C. (2014). Novel mechanisms of endothelial mechanotransduction. *Arterioscler Thromb Vasc Biol*, **34**, pp. 2378–2386.
8. Baeyens, N., Bandyopadhyay, C., Coon, B. G., et al. (2016). Endothelial fluid shear stress sensing in vascular health and disease. *J Clin Invest*, **126**, pp. 821–828.
9. Rajendran, P., Rengarajan, T., Thangavel, J., et al. (2013). The vascular endothelium and human diseases. *Int J Biol Sci*, **9**, pp. 1057–1069.
10. Fledderus, J. O., van Thienen, J. V., Boon, R. A., et al. (2007). Prolonged shear stress and KLF2 suppress constitutive proinflammatory transcription through inhibition of ATF2. *Blood*, **109**, pp. 4249–4257.

11. Parmar, K. M., Larman, H. B., Dai, G., et al. (2006). Integration of flow-dependent endothelial phenotypes by Kruppel-like factor 2. *J Clin Invest*, **116**, pp. 49–58.
12. Wang, N., Miao, H., Li, Y. S., et al. (2006). Shear stress regulation of Kruppel-like factor 2 expression is flow pattern-specific. *Biochem Biophys Res Commun*, **341**, pp. 1244–1251.
13. Hamik, A., Lin, Z., Kumar, A., et al. (2007). Kruppel-like factor 4 regulates endothelial inflammation. *J Biol Chem*, **282**, pp. 13769–13779.
14. Villarreal, G., Jr., Zhang, Y., Larman, H. B., et al. (2010). Defining the regulation of KLF4 expression and its downstream transcriptional targets in vascular endothelial cells. *Biochem Biophys Res Commun*, **391**, pp. 984–989.
15. Clark, P. R., Jensen, T. J., Kluger, M. S., et al. (2011). MEK5 is activated by shear stress, activates ERK5 and induces KLF4 to modulate TNF responses in human dermal microvascular endothelial cells. *Microcirculation*, **18**, pp. 102–117.
16. Hosoya, T., Maruyama, A., Kang, M. I., et al. (2005). Differential responses of the Nrf2-Keap1 system to laminar and oscillatory shear stresses in endothelial cells. *J Biol Chem*, **280**, pp. 27244–27250.
17. Partridge, J., Carlsen, H., Enesa, K., et al. (2007). Laminar shear stress acts as a switch to regulate divergent functions of NF-kappaB in endothelial cells. *FASEB J*, **21**, pp. 3553–3561.
18. Xu, S., Koroleva, M., Yin, M., et al. (2016). Atheroprotective laminar flow inhibits Hippo pathway effector YAP in endothelial cells. *Transl Res*, **176**, pp. 18–28 e2.
19. Wang, L., Luo, J. Y., Li, B., et al. (2016). Integrin-YAP/TAZ-JNK cascade mediates atheroprotective effect of unidirectional shear flow. *Nature*, **540**(7634), pp. 579–582.
20. Wang, K. C., Yeh, Y. T., Nguyen, P., et al. (2016). Flow-dependent YAP/TAZ activities regulate endothelial phenotypes and atherosclerosis. *Proc Natl Acad Sci U S A*, **113**, pp. 11525–11530.
21. Feng, S., Bowden, N., Fragiadaki, M., et al. (2017). Mechanical activation of hypoxia-inducible factor 1alpha drives endothelial dysfunction at atheroprone sites. *Arterioscler Thromb Vasc Biol*, **37**, pp. 2087–2101.
22. Wu, D., Huang, R. T., Hamanaka, R. B., et al. (2017). HIF-1alpha is required for disturbed flow-induced metabolic reprogramming in human and porcine vascular endothelium. *Elife*, **6**, pp. e25217.

23. Maejima, T., Inoue, T., Kanki, Y., et al. (2014). Direct evidence for pitavastatin induced chromatin structure change in the KLF4 gene in endothelial cells. *PLoS One*, **9**, pp. e96005.

24. Parmar, K. M., Nambudiri, V., Dai, G., et al. (2005). Statins exert endothelial atheroprotective effects via the KLF2 transcription factor. *J Biol Chem*, **280**, pp. 26714–26719.

25. Marrone, G., Russo, L., Rosado, E., et al. (2013). The transcription factor KLF2 mediates hepatic endothelial protection and paracrine endothelial-stellate cell deactivation induced by statins. *J Hepatol*, **58**, pp. 98–103.

26. Gracia-Sancho, J., Villarreal, G., Jr., Zhang, Y., et al. (2010). Activation of SIRT1 by resveratrol induces KLF2 expression conferring an endothelial vasoprotective phenotype. *Cardiovasc Res*, **85**, pp. 514–519.

27. Xu, Y., Liu, P., Xu, S., et al. (2017). Tannic acid as a plant-derived polyphenol exerts vasoprotection via enhancing KLF2 expression in endothelial cells. *Sci Rep*, **7**, pp. 6686.

28. Xu, Y., Xu, S., Liu, P., et al. (2017). Suberanilohydroxamic acid as a pharmacological Kruppel-like factor 2 activator that represses vascular inflammation and atherosclerosis. *J Am Heart Assoc*, **6**, pp. e007134.

29. Yan, Z., Deng, Y., Jiao, F., et al. (2017). Lipopolysaccharide downregulates Kruppel-like factor 2 (KLF2) via inducing DNMT1-mediated hypermethylation in endothelial cells. *Inflammation*, **40**, pp. 1589–1598.

30. Kumar, A., Lin, Z., SenBanerjee, S., et al. (2005). Tumor necrosis factor alpha-mediated reduction of KLF2 is due to inhibition of MEF2 by NF-kappaB and histone deacetylases. *Mol Cell Biol*, **25**, pp. 5893–5903.

31. Berk, B. C. (2008). Atheroprotective signaling mechanisms activated by steady laminar flow in endothelial cells. *Circulation*, **117**, pp. 1082–1089.

32. Deng, Q., Huo, Y. and Luo, J. (2014). Endothelial mechanosensors: the gatekeepers of vascular homeostasis and adaptation under mechanical stress. *Sci China Life Sci*, **57**, pp. 755–762.

33. Givens, C. and Tzima, E. (2016). Endothelial mechanosignaling: does one sensor fit all? *Antioxid Redox Signal*, **25**, pp. 373–388.

34. Zaragoza, C., Marquez, S. and Saura, M. (2012). Endothelial mechanosensors of shear stress as regulators of atherogenesis. *Curr Opin Lipidol*, **23**, pp. 446–452.

35. Fleming, I., Fisslthaler, B., Dixit, M., et al. (2005). Role of PECAM-1 in the shear-stress-induced activation of Akt and the endothelial nitric oxide synthase (eNOS) in endothelial cells. *J Cell Sci*, **118**, pp. 4103–4111.

36. Xu, S., Ha, C. H., Wang, W., et al. (2016). PECAM1 regulates flow-mediated Gab1 tyrosine phosphorylation and signaling. *Cell Signal*, **28**, pp. 117–124.

37. Conway, D. E., Coon, B. G., Budatha, M., et al. (2017). VE-cadherin phosphorylation regulates endothelial fluid shear stress responses through the polarity protein LGN. *Curr Biol*, **27**, pp. 2219–2225 e5.

38. Tzima, E., Irani-Tehrani, M., Kiosses, W. B., et al. (2005). A mechanosensory complex that mediates the endothelial cell response to fluid shear stress. *Nature*, **437**, pp. 426–431.

39. Coon, B. G., Baeyens, N., Han, J., et al. (2015). Intramembrane binding of VE-cadherin to VEGFR2 and VEGFR3 assembles the endothelial mechanosensory complex. *J Cell Biol*, **208**, pp. 975–986.

40. Baeyens, N., Nicoli, S., Coon, B. G., et al. (2015). Vascular remodeling is governed by a VEGFR3-dependent fluid shear stress set point. *Elife*, **4**, pp. e04645.

41. Harry, B. L., Sanders, J. M., Feaver, R. E., et al. (2008). Endothelial cell PECAM-1 promotes atherosclerotic lesions in areas of disturbed flow in ApoE-deficient mice. *Arterioscler Thromb Vasc Biol*, **28**, pp. 2003–2008.

42. Goel, R., Schrank, B. R., Arora, S., et al. (2008). Site-specific effects of PECAM-1 on atherosclerosis in LDL receptor-deficient mice. *Arterioscler Thromb Vasc Biol*, **28**, pp. 1996–2002.

43. Mendoza, S. A., Fang, J., Gutterman, D. D., et al. (2010). TRPV4-mediated endothelial Ca^{2+} influx and vasodilation in response to shear stress. *Am J Physiol Heart Circ Physiol*, **298**, pp. H466– H476.

44. Ye, L., Xu, M., Hu, M., et al. (2018). TRPV4 is involved in irisin-induced endothelium-dependent vasodilation. *Biochem Biophys Res Commun*, **495**, pp. 41–45.

45. He, D., Pan, Q., Chen, Z., et al. (2017). Treatment of hypertension by increasing impaired endothelial TRPV4-KCa2.3 interaction. *EMBO Mol Med*, **9**, pp. 1491–1503.

46. Caires, R., Sierra-Valdez, F. J., Millet, J. R. M., et al. (2017). Omega-3 fatty acids modulate TRPV4 function through plasma membrane remodeling. *Cell Rep*, **21**, pp. 246–258.

47. Naik, J. S., Osmond, J. M., Walker, B. R., et al. (2016). Hydrogen sulfide-induced vasodilation mediated by endothelial TRPV4 channels. *Am J Physiol Heart Circ Physiol*, **311**, pp. H1437–H1444.

48. Baratchi, S., Almazi, J. G., Darby, W., et al. (2016). Shear stress mediates exocytosis of functional TRPV4 channels in endothelial cells. *Cell Mol Life Sci*, **73**, pp. 649–666.

49. Xue, C., Gong, J., Guo, Y., et al. (2017). Oxygenized low density lipoprotein down-regulates the TRPV4 protein expression of macrophage through activation of peroxisome proliferator-activated receptor gamma. *Minerva Med*, **108**, pp. 1–12.

50. Goswami, R., Merth, M., Sharma, S., et al. (2017). TRPV4 calcium-permeable channel is a novel regulator of oxidized LDL-induced macrophage foam cell formation. *Free Radic Biol Med*, **110**, pp. 142–150.

51. Xu, S., Liu, B., Yin, M., et al. (2016). A novel TRPV4-specific agonist inhibits monocyte adhesion and atherosclerosis. *Oncotarget*, **7**, pp. 37622–37635.

52. Beech, D. J. (2017). Endothelial Piezo1 channels as sensors of exercise. *J Physiol*, **596**, pp. 979–984.

53. Bagriantsev, S. N., Gracheva, E. O. and Gallagher, P. G. (2014). Piezo proteins: regulators of mechanosensation and other cellular processes. *J Biol Chem*, **289**, pp. 31673–31681.

54. Zhang, T., Chi, S., Jiang, F., et al. (2017). A protein interaction mechanism for suppressing the mechanosensitive Piezo channels. *Nat Commun*, **8**, pp. 1797.

55. Ranade, S. S., Qiu, Z., Woo, S. H., et al. (2014). Piezo1, a mechanically activated ion channel, is required for vascular development in mice. *Proc Natl Acad Sci U S A*, **111**, pp. 10347–10352.

56. Li, J., Hou, B., Tumova, S., et al. (2014). Piezo1 integration of vascular architecture with physiological force. *Nature*, **515**, pp. 279–282.

57. Syeda, R., Xu, J., Dubin, A. E., et al. (2015). Chemical activation of the mechanotransduction channel Piezo1. *Elife*, **4**, pp. 07369.

58. Wang, S., Chennupati, R., Kaur, H., et al. (2016). Endothelial cation channel PIEZO1 controls blood pressure by mediating flow-induced ATP release. *J Clin Invest*, **126**, pp. 4527–4536.

59. Dela Paz, N. G. and Frangos, J. A. (2018). Yoda1-induced phosphorylation of Akt and ERK1/2 does not require Piezo1 activation. *Biochem Biophys Res Commun*, **497**, pp. 220–225.

60. Iomini, C., Tejada, K., Mo, W., et al. (2004). Primary cilia of human endothelial cells disassemble under laminar shear stress. *J Cell Biol*, **164**, pp. 811–817.

61. Nauli, S. M., Kawanabe, Y., Kaminski, J. J., et al. (2008). Endothelial cilia are fluid shear sensors that regulate calcium signaling and nitric oxide production through polycystin-1. *Circulation*, **117**, pp. 1161–1171.

62. Van der Heiden, K., Hierck, B. P., Krams, R., et al. (2008). Endothelial primary cilia in areas of disturbed flow are at the base of atherosclerosis. *Atherosclerosis*, **196**, pp. 542–550.

63. Hierck, B. P., Van der Heiden, K., Alkemade, F. E., et al. (2008). Primary cilia sensitize endothelial cells for fluid shear stress. *Dev Dyn*, **237**, pp. 725–735.

64. Dinsmore, C. and Reiter, J. F. (2016). Endothelial primary cilia inhibit atherosclerosis. *EMBO Rep*, **17**, pp. 156–166.

65. Sanchez-Duffhues, G., de Vinuesa, A. G., Lindeman, J. H., et al. (2015). SLUG is expressed in endothelial cells lacking primary cilia to promote cellular calcification. *Arterioscler Thromb Vasc Biol*, **35**, pp. 616–627.

66. Sinha, B., Koster, D., Ruez, R., et al. (2011). Cells respond to mechanical stress by rapid disassembly of caveolae, *Cell*, **144**, pp. 402–413.

67. Boyd, N. L., Park, H., Yi, H., et al. (2003). Chronic shear induces caveolae formation and alters ERK and Akt responses in endothelial cells. *Am J Physiol Heart Circ Physiol*, **285**, pp. H1113– H1122.

68. Chai, Q., Wang, X. L., Zeldin, D. C., et al. (2013). Role of caveolae in shear stress-mediated endothelium-dependent dilation in coronary arteries. *Cardiovasc Res*, **100**, pp. 151–159.

69. Yu, J., Bergaya, S., Murata, T., et al. (2006). Direct evidence for the role of caveolin-1 and caveolae in mechanotransduction and remodeling of blood vessels. *J Clin Invest*, **116**, pp. 1284–1291.

70. Engel, D., Beckers, L., Wijnands, E., et al. (2011). Caveolin-1 deficiency decreases atherosclerosis by hampering leukocyte influx into the arterial wall and generating a regulatory T-cell response. *FASEB J*, **25**, pp. 3838–3848.

71. Fernandez-Hernando, C., Yu, J., Davalos, A., et al. (2010). Endothelial-specific overexpression of caveolin-1 accelerates atherosclerosis in apolipoprotein E-deficient mice. *Am J Pathol*, **177**, pp. 998–1003.

72. Martinez, P., Thanasoula, M., Carlos, A. R., et al. (2010). Mammalian Rap1 controls telomere function and gene expression through binding to telomeric and extratelomeric sites. *Nat Cell Biol*, **12**, pp. 768–780.

73. Chrzanowska-Wodnicka, M. (2017). Rap1 in endothelial biology. *Curr Opin Hematol*, **24**, pp. 248–255.
74. Lakshmikanthan, S., Zheng, X., Nishijima, Y., et al. (2015). Rap1 promotes endothelial mechanosensing complex formation, NO release and normal endothelial function. *EMBO Rep*, **16**, pp. 628–637.
75. Wong, K. H. K., Cai, Y., Ying, F., et al. (2018). Deletion of Rap1 disrupts redox balance and impairs endothelium-dependent relaxations. *J Mol Cell Cardiol*, **115**, pp. 1–9.
76. Ramos, C. J., Lin, C., Liu, X., et al. (2018). The EPAC-Rap1 pathway prevents and reverses cytokine-induced retinal vascular permeability. *J Biol Chem*, **293**, pp. 717–730.
77. Cai, Y., Sukhova, G. K., Wong, H. K., et al. (2015). Rap1 induces cytokine production in pro-inflammatory macrophages through NFkappaB signaling and is highly expressed in human atherosclerotic lesions. *Cell Cycle*, **14**, pp. 3580–3592.
78. Zeng, Y. and Tarbell, J. M. (2014). The adaptive remodeling of endothelial glycocalyx in response to fluid shear stress. *PLoS One*, **9**, pp. e86249.
79. Tarbell, J. M. and Cancel, L. M. (2016). The glycocalyx and its significance in human medicine. *J Intern Med*, **280**, pp. 97–113.
80. Baeyens, N., Mulligan-Kehoe, M. J., Corti, F., et al. (2014). Syndecan 4 is required for endothelial alignment in flow and atheroprotective signaling. *Proc Natl Acad Sci U S A*, **111**, pp. 17308–17313.
81. Ebong, E. E., Lopez-Quintero, S. V., Rizzo, V., et al. (2014). Shear-induced endothelial NOS activation and remodeling via heparan sulfate, glypican-1, and syndecan-1, *Integr Biol (Camb)*, **6**, pp. 338–347.
82. Lu, T., Wang, X. L., Chai, Q., et al. (2017). Role of the endothelial caveolae microdomain in shear stress-mediated coronary vasorelaxation. *J Biol Chem*, **292**, pp. 19013–19023.
83. Yao, Y., Rabodzey, A. and Dewey, C. F., Jr. (2007). Glycocalyx modulates the motility and proliferative response of vascular endothelium to fluid shear stress. *Am J Physiol Heart Circ Physiol*, **293**, pp. H1023– H1030.
84. Alphonsus, C. S. and Rodseth, R. N. (2014). The endothelial glycocalyx: a review of the vascular barrier. *Anaesthesia*, **69**, pp. 777–784.
85. Ross, T. D., Coon, B. G., Yun, S., et al. (2013). Integrins in mechanotransduction. *Curr Opin Cell Biol*, **25**, pp. 613–618.
86. Shyy, J. Y. and Chien, S. (2002). Role of integrins in endothelial mechanosensing of shear stress. *Circ Res*, **91**, pp. 769–775.

87. Chen, J., Green, J., Yurdagul, A., Jr., et al. (2015). alphavbeta3 Integrins mediate flow-induced NF-kappaB activation, proinflammatory gene expression, and early atherogenic inflammation. *Am J Pathol*, **185**, pp. 2575–2589.

88. Orr, A. W., Ginsberg, M. H., Shattil, S. J., et al. (2006). Matrix-specific suppression of integrin activation in shear stress signaling. *Mol Biol Cell*, **17**, pp. 4686–4697.

89. Yang, T. L., Lee, P. L., Lee, D. Y., et al. (2018). Differential regulations of fibronectin and laminin in Smad2 activation in vascular endothelial cells in response to disturbed flow. *J Biomed Sci*, **25**, pp. 1.

90. Hu, S., Liu, Y., You, T., et al. (2018). Vascular semaphorin 7A upregulation by disturbed flow promotes atherosclerosis through endothelial beta1 integrin. *Arterioscler Thromb Vasc Biol*, **38**, pp. 335–343.

91. Yang, B. and Rizzo, V. (2013). Shear stress activates eNOS at the endothelial apical surface through beta1 containing integrins and caveolae. *Cell Mol Bioeng*, **6**, pp. 346–354.

92. Strohbach, A., Pennewitz, M., Glaubitz, M., et al. (2018). The apelin receptor influences biomechanical and morphological properties of endothelial cells. *J Cell Physiol*, doi: 10.1002/jcp.26496.

93. Otte, L. A., Bell, K. S., Loufrani, L., et al. (2009). Rapid changes in shear stress induce dissociation of a G alpha(q/11)-platelet endothelial cell adhesion molecule-1 complex. *J Physiol*, **587**, pp. 2365–2373.

94. dela Paz, N. G., Melchior, B., Shayo, F. Y., et al. (2014). Heparan sulfates mediate the interaction between platelet endothelial cell adhesion molecule-1 (PECAM-1) and the Galphaq/11 subunits of heterotrimeric G proteins. *J Biol Chem*, **289**, pp. 7413–7424.

95. Dela Paz, N. G., Melchior, B. and Frangos, J. A. (2017). Shear stress induces Galphaq/11 activation independently of G protein-coupled receptor activation in endothelial cells. *Am J Physiol Cell Physiol*, **312**, pp. C428–C437.

96. Mack, J. J., Mosqueiro, T. S., Archer, B. J., et al. (2017). NOTCH1 is a mechanosensor in adult arteries. *Nat Commun*, **8**, pp. 1620.

97. Sawamura, T., Kume, N., Aoyama, T., et al. (1997). An endothelial receptor for oxidized low-density lipoprotein. *Nature*, **386**, pp. 73–77.

98. Xu, S., Ogura, S., Chen, J., et al. (2013). LOX-1 in atherosclerosis: biological functions and pharmacological modifiers. *Cell Mol Life Sci*, **70**, pp. 2859–2872.

99. Lee, J. Y., Chung, J., Kim, K. H., et al. (2018). Fluid shear stress regulates the expression of Lectin-like oxidized low density lipoprotein

receptor-1 via KLF2-AP-1 pathway depending on its intensity and pattern in endothelial cells. *Atherosclerosis*, **270**, pp. 76–88.

100. Hansson, G. K. (2005). Inflammation, atherosclerosis, and coronary artery disease. *N Engl J Med*, **352**, pp. 1685–1695.
101. Wierda, R. J., Geutskens, S. B., Jukema, J. W., et al. (2010). Epigenetics in atherosclerosis and inflammation. *J Cell Mol Med*, **14**, pp. 1225–1240.
102. Xu, S., Pelisek, J. and Jin, Z. G. (2018). Atherosclerosis is an epigenetic disease. *Trends Endocrinol Metab*, **29**(11), pp. 739–742.
103. Fang, J., Little, P. J. and Xu, S. (2018). Atheroprotective effects and molecular targets of tanshinones derived from herbal medicine Danshen. *Med Res Rev*, **38**, pp. 201–228.
104. Xu, S., Bai, P., Little, P. J., et al. (2014). Poly(ADP-ribose) polymerase 1 (PARP1) in atherosclerosis: from molecular mechanisms to therapeutic implications. *Med Res Rev*, **34**, pp. 644–675.
105. Xu, S., Xu, Y., Yin, M., et al. (2018). Flow-dependent epigenetic regulation of IGFBP5 expression by H3K27me3 contributes to endothelial anti-inflammatory effects. *Theranostics*, **8**, pp. 3007–3021.
106. Ferguson, J. F., Allayee, H., Gerszten, R. E., et al. (2016). Nutrigenomics, the microbiome, and gene-environment interactions: new directions in cardiovascular disease research, prevention, and treatment: a scientific statement from the American Heart Association. *Circ Cardiovasc Genet*, **9**, pp. 291–313.
107. Flowers, E., Froelicher, E. S. and Aouizerat, B. E. (2012). Gene-environment interactions in cardiovascular disease. *Eur J Cardiovasc Nurs*, **11**, pp. 472–478.
108. Xu, J., Mathur, J., Vessieres, E., et al. (2018). GPR68 senses flow and is essential for vascular physiology. *Cell*, **173**, pp. 762–775.e16.

Chapter 2

Role of Krüppel-Like Factors in Endothelial Cell Function and Shear Stress–Mediated Vasoprotection

Eugene Chang and Mukesh Jain
Case Cardiovascular Research Institute, Wolstein Research Building, Room 4-405, 2103 Cornell Road, Cleveland, OH 4410, USA
Eugene.Chang@case.edu; Mukesh.Jain@HarringtonDiscovery.org

Laminar shear stress is an important regulator of endothelial biology. Endothelial cells transduce the biomechanical frictional forces from blood flow to regulate gene expression and cell phenotype through numerous signal transduction cascades. Krüppel-like factors (KLFs) are a family of zinc-finger-type transcription factors that play important roles in modulating cellular function in a broad range of mammalian cell types. Several KLFs have been reported to be expressed in endothelial cells. Of these, KLF2 and KLF4 have been shown to be upregulated by laminar shear stress and serve as master regulators of mediating the vasoprotective functions of shear stress against vascular disease, including atherogenesis and thrombosis. This chapter will review the essential functions of KLFs

Modern Mechanobiology: Convergence of Biomechanics, Development, and Genomics
Edited by Juhyun Lee, Sharon Gerecht, Hanjoong Jo, and Tzung Hsiai

ISBN 978-981-4800-58-7 (Hardcover), 978-0-429-29483-9 (eBook)
www.jennystanford.com

in endothelial cell biology, with a focus on their role in laminar shear stress–mediated vasoprotection.

2.1 Introduction

Cardiovascular disease is the leading cause of morbidity and mortality in the world and accounted for nearly one-third of all deaths globally in 2015 [1, 2]. The accumulation of plaque and damage to the vasculature through atherosclerosis is the primary pathological mechanism of cardiovascular disease [3]. Dysfunction of the endothelial cells that line the vasculature is responsible for the pathological phenotype seen in atherosclerosis, including dysregulation of vasodilation, barrier function, inflammation, and thrombosis [4]. Although the development of atherosclerosis is dependent upon a complex interplay between many factors and processes, there is now clear basic and clinical evidence that shear stress plays a central role in atherosclerosis [5]. It has been well established that there is a non-uniform distribution of atherosclerosis within the vasculature [6, 7]. Atherosclerosis occurs more frequently at branch points of the arterial tree—areas that are exposed to turbulent or disturbed blood flow. In contrast, laminar shear stress is present in unbranched portions of vessels and induces anti-inflammatory and anticoagulant genes, such as endothelial nitric oxide synthase (*eNOS*) and thrombomodulin (*TM*), thereby conferring an atheroprotective phenotype to the vessel wall.

Studies from our group and others have demonstrated a critical role of the Krüppel-like factor (KLF) family of zinc-finger transcription factors in endothelial biology. KLF2 and KLF4 are upregulated by laminar shear stress and have been shown to mediate the vasoprotective effects of laminar shear stress in endothelial cells. This chapter describes our current understanding of the role of KLFs in the shear stress–mediated endothelial phenotype.

2.2 Krüppel-Like Factors

The Sp/KLF family of transcription factors is a subclass of the zinc-finger family of transcriptional regulators that broadly regulate cellular growth and differentiation [8–11]. KLFs exhibit homology to the *Drosophila melanogaster* segmentation gene product Krüppel. Krüppel is the German word for "cripple"; the name is derived from the observation that mutation of this protein causes deletion of thoracic and anterior abdominal segments in *Drosophila* embryos [12–14]. The first mammalian KLF to be discovered, KLF1 or erythroid KLF (EKLF), was cloned from red blood cells in 1993 by Bieker and colleagues [15]. There are currently 18 mammalian KLFs (KLF1–18) that have been identified and are expressed in various tissues with distinct expression patterns and functions [16–18]. Members of the KLF family share three highly conserved zinc-finger regions in the C-terminal that bind with varying affinities to a consensus DNA sequence (termed "GC-box element" or "CACCC element") and exert diverse transcriptional functions [10, 19]. The transactivation and transrepression domains are in the *N*-terminal and provide numerous potential sites for regulation via unique post-translational modification sites and protein interaction motifs [19, 20]. Furthermore, KLFs can modulate other members of their family through several distinct mechanisms, such as through cross regulation of expression or direct protein-protein interaction [10, 16, 18, 21, 22].

In endothelial cells, KLF2, KLF4, KLF6, and KLF15 play major roles in endothelial biology, with KLF2 and KLF4 being the most important in mediating shear stress vasoprotection. Cellular processes critical to maintaining a healthy endothelium, including barrier function, inflammation, coagulation, vascular tone, vasodilation, and angiogenesis, are all regulated by the KLFs [18, 19]. Next, we will discuss the endothelial KLFs and their roles in human disease.

2.2.1 Krüppel-Like Factor 2

KLF2 was first isolated and cloned by Lingrel and colleagues in 1995 using the zinc-finger domain of EKLF as a hybridization probe. The full-length mouse KLF2 cDNA encodes a protein of 354 amino acids (355 amino acids in humans) and a molecular weight of 38 kDa.

The human and mouse KLF2 homologs exhibit an 85% nucleotide identity and 90% amino acid similarity. Within the 5' proximal promoter of KLF2, a region of 75 nucleotides appears particularly important in transcriptional regulation of KLF2. Mutagenesis of this region significantly impairs the ability of the KLF2 promoter to drive reporter gene expression [23].

KLF2 has both activating and inhibitory effects on gene transcription via both direct and indirect promoter activity. Transcriptional regulation by KLF2 is mediated by a potent transcriptional activation domain (amino acids 1–110) as well as an autoinhibitory domain encoded by amino acids 111–267 [24]. An interesting function of the inhibitory domain is its ability to mediate polyubiquitination and proteasomal degradation of KLF2 through binding with the WW domain–containing E3 ubiquitin protein ligase 1 [25].

The developmental expression of KLF2 is highly regulated, with initial expression at embryonic day 7 (E7), deactivation at E11, and subsequent reactivation at E15 [26]. KLF2 is expressed in multiple cell types and has been implicated in endothelial biology [27–31], monocyte/macrophage biology [32, 33], T-cell biology [34–37], erythropoiesis [38], adipocyte biology [39, 40], and stem cell biology [41]. KLF2 is highly expressed in the lung and thus was initially termed "lung Krüppel-like factor" [26]. In fact, KLF2 is vital for normal lung development, as demonstrated by chimeric KLF2-null mice. These mice fail to develop normal lung tissue, whereas other tissues are normally developed [23]. KLF2 knockout mice have a grossly normal vascular network, with vasculogenesis and angiogenesis seemingly intact [42, 43]. However, as will be described in detail below, vessel maturation is impaired, leading to embryonic death from hemorrhage. Thus, KLF2 has been shown to have a role in multiple cell types and a crucial function in lung and vascular development. A schematic diagram of the regulation of KLF2 in endothelial cells is shown in Fig. 2.1.

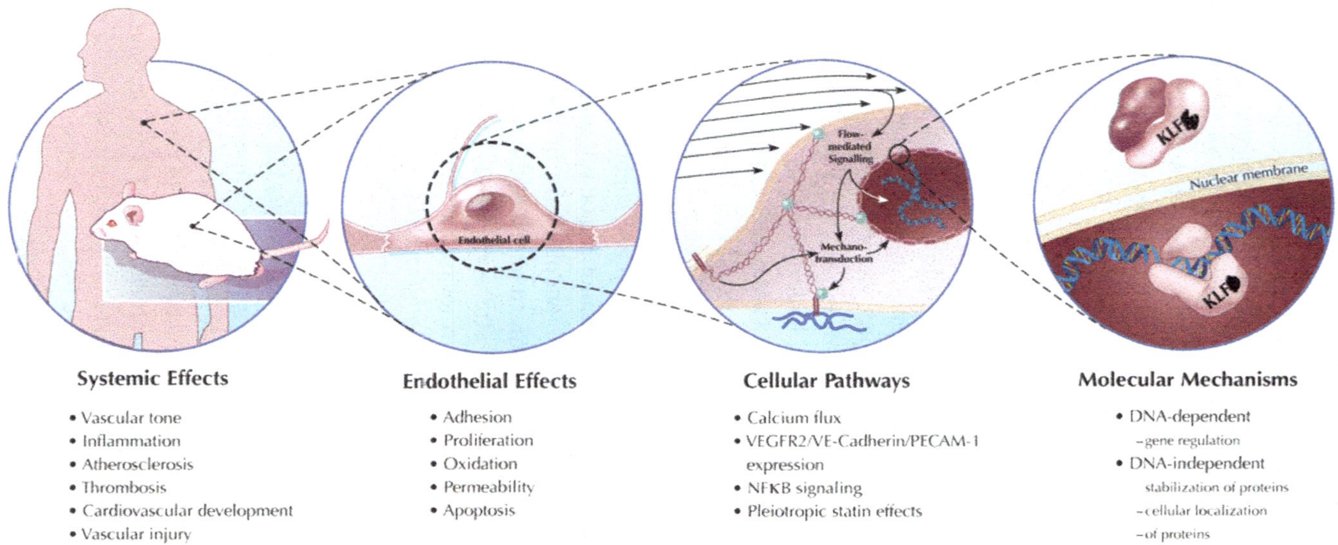

Figure 2.1 Schematic diagram of the regulation of KLF2 in endothelial cells. Ang-1, angiopoietin 1; AGE, advanced glycation end products.

2.2.1.1 Regulation of KLF2 by laminar shear stress

The susceptibility of vascular branch points and curvatures to atherosclerosis is in large part related to exposure to turbulent or disturbed flow. In contrast, the portion of the arterial tree exposed to unidirectional laminar shear stress tends to be protected. Mechanotransduction of these hemodynamic forces by endothelial cells activates numerous downstream pathways necessary for vascular homeostasis [5]. Many of these signal transduction pathways converge on the KLFs, making them master regulators of the vasoprotective shear stress–induced endothelial phenotype. KLF2 and KLF4 are strongly induced by laminar flow through a mechanotransduction pathway on the endothelial surface consisting of platelet endothelial cell adhesion molecule-1 (PECAM1), vascular endothelial (VE)-cadherin, and VE growth factor receptor 2/3 (VEGFR2/3). This leads to downstream activation of the mitogen/extracellular signal–regulated kinase kinase-5, extracellular signal-regulated kinase-5, myocyte enhancing factor 2 (MEK5/ERK5/MEF2) pathway that transregulates KLF [5, 29, 44].

This mechanism was first elucidated in 2002 by Horrevoets and colleagues, who used microarray analysis to identify endothelial KLF2 as induced by prolonged laminar shear stress [45]. In situ hybridization assays in the human aorta revealed that KLF2 is consistently expressed in the endothelium of the aorta but its expression level is lower at branch points of the aortic arch and the abdominal aorta [46]. Shear stress–mediated KLF2 induction has also been observed during cardiovascular development in chicken embryos. Groenendijk et al. demonstrated the overlap of KLF2's expression pattern with the distribution of high shear levels, which occurs in the narrow regions of the cardiovascular system, such as the cardiac inflow/outflow tracts and the atrioventricular canal, and in the early stages of development, in the aortic sac and the pharyngeal arch arteries [47, 48]. Using a murine carotid artery collar model to focally increase flow, Dekker et al. demonstrated the direct involvement of high shear stress in the in vivo induction of KLF2 expression in the endothelium [46]. Laminar shear stress induces KLF2 expression by increasing both transcription RNA and messenger RNA (mRNA) stability [45, 49]. Furthermore, alignment of actin cytoskeletal fibers in endothelial cells in response to shear stress is regulated by KLF2 through its inhibition of Jun NH_2-terminal kinase and its downstream targets ATF2/c-jun [50].

Lingrel and colleagues identified a novel 62 bp shear stress responsive element in the KLF2 promoter that contains a 30 bp tripartite palindrome motif [51,52]. Furthermore, they demonstrated binding of p300/cAMP-response element-binding protein-binding protein-associated factor (PCAF) and heterogenous nuclear ribonucleoprotein D (hnRNP D) to this region as components of a shear stress–specific regulatory complex. Binding of these factors was dependent upon the activation of phosphatidylinositol 3-kinase (PI3K) pathway and correlated with histone H3 and H4 acetylation [51]. In a subsequent study, this group identified nucleolin, an abundant and ubiquitous cellular protein that plays important roles in DNA transcription, chromatin remodeling, and mRNA stability [53], as a factor that binds the shear stress–responsive site of the KLF2 promoter. The binding of nucleolin is PI3K dependent and required for the induction of KLF2 by laminar shear stress [54]. Wang and colleagues reported that the Src signaling pathway is also involved in the shear stress–mediated regulation of KLF2. They demonstrated that Src mediates the inhibitory effect of oscillatory or disturbed flow on KLF2 expression [55].

Additional investigation of the mechanism of flow-induced KLF2 expression has demonstrated that the MEK5/ERK5/MEF2 signaling pathway mediates the increase of KLF2 by laminar shear stress. Kumar and colleagues showed that a single consensus MEF2 binding site in the conserved region of the KLF2 promoter regulates KLF2 promoter activity [28]. ERK5 (also referred to as big mitogen-activated protein kinase-1 [MAPK1] or BMK1) is highly induced by laminar shear stress, and MEF2 is a well-characterized target gene of ERK5. Winoto and colleagues found that ERK5 plays a crucial role in KLF2 expression in embryos and that MEF2 mediates ERK5-induced KLF2 expression [56]. Building on these observations, Parmar et al. demonstrated that dominant-negative MEF2 or mutant MEK5 (an upstream activator of ERK5) abolished flow-mediated KLF2 induction in endothelial cells [57]. Young et al. demonstrated that the shear stress–regulated factor AMP-activated protein kinase is also able to activate the ERK5/MEF2 pathway, thereby increasing

KLF2 expression [58]. Woo and colleagues demonstrated that SUMOylation of ERK5 by H_2O_2 and advanced glycation end products (AGEs), major transducers of diabetic vasculopathy, inhibits shear stress–mediated KLF2 induction in endothelial cells [59]. SUMOylation is a post-translational modification in which small ubiquitin-like modifier (SUMO) proteins are attached to modify target function. They observed that H_2O_2 and AGEs induce endogenous SUMOylation of ERK5, leading to the reduction of shear stress–induced eNOS expression. SUMOylation of ERK5 downregulates MEF2 and subsequently reduces KLF2 transcriptional activity, thereby suppressing induction of eNOS and the anti-inflammatory response by shear stress [59]. It has been shown that angiopoeitin-1 (Ang-1) also upregulates KLF2 expression through the PI3K/AKT-dependent activation of MEF2 [60].

The adaptor protein p66shc, known to promote cellular oxidative stress and apoptosis, has been demonstrated to inhibit KFL2 transcription by suppression of MEF2A expression [61]. These observations indicate that the ERK5/MEF2 signaling pathway plays a crucial role in regulating KLF2 expression by laminar shear stress.

The identification of MEF2 as being critical in the regulation of KLF2 also has implications for the downregulation of KLF2 expression observed in the setting of cytokine stimulation. Specifically, Kumar et al. showed that tumor necrosis factor-α (TNF-α) downregulates KLF2 expression in endothelial cells. Inhibition of nuclear factor kappa light chain enhancer of activated B cells (NF-κB) abolishes the reduction of KLF2 by TNF-α as supported by the inability of TNF-α to downregulate KLF2 in p65 null cells. Mechanistic studies have revealed that histone deacetylase (HDAC)-4/5 and p65 form a complex and binding of this complex to MEF2 site of the KLF2 promoter causes TNF-α-mediated KLF2 repression [28]. This is supported by other studies that show that shear stress stimulates a phosphorylation-dependent nuclear export of HDAC-5, further enhancing MEF2 transactivation and KLF2 expression [62, 63]. Areas of disturbed flow have elevated the expression of HDAC-3/5/7, and this disturbed flow induces the association of these HDACs with MEF2, leading to KLF2 transrepression [64].

2.2.1.2 Targets of shear stress–induced KLF2

2.2.1.2.1 *Inflammation and atherogenesis*

Of particular importance is the ability of shear-induced KLF2 to mediate the favorable effects of shear stress. Indeed, more than 15% of all flow-regulated genes are dependent on flow-mediated KLF2 induction, and of those most highly regulated by flow, approximately 50% are KLF2 dependent [57]. The Horrevoetts laboratory has suggested that three characteristics of KLF2—(i) potent upregulation by prolonged laminar shear stress, (ii) lack of upregulation by other physiological mediators, and (iii) vascular expression limited to the endothelium—make it an attractive candidate as an integral mediator of laminar flow [45]. Targets of KLF2 (direct or indirect) are involved in mediating a broad range of shear-regulated systemic, cellular, and molecular events. The effect of KLF2 on regulation of the expression of two critical mediators of vascular tone, eNOS and endothelin-1 (ET-1), has been published by several laboratories [29, 46, 57]. As described in detail in earlier chapters, nitric oxide (NO) production in response to shear stress is a fundamental mechanism of regulation of vascular tone. KLF2 is one of the most potent inducers of eNOS expression yet described [29]. Mechanistic studies have revealed that KLF2 directly binds to the eNOS promoter. Furthermore, KLF2 recruits the coactivator CREB-binding protein (CBP)/p300 to the eNOS promoter [29]. KLF2 has also been shown to induce C-natriuretic peptide and arginosuccinate synthase, a limiting enzyme in eNOS substrate bioavailability [44, 46, 65]. Finally, KLF2 reduces the expression of caveolin-1, which negatively regulates eNOS activity [66]. In concert, KLF2 inhibits the expression of adrenomedullin, ET-1, and angiotensin-converting enzyme, all of which increase vascular tone [46]. Surprisingly, the in vivo phenotype of conditional embryonic loss of systemic or endothelial-specific KLF2 is lethal embryonic heart failure, thought to be secondary to low vascular tone [67]. The phenotype, demonstrated in both mice and zebrafish, is rescued by the administration of phenylephrine. A molecular mechanism for the phenotype could not be demonstrated in this study as the expressions of target genes known to affect vascular tone (including eNOS, adrenomedullin, and ET-1, as well as smooth muscle cell function) were similar in the controls and knockout mice. The authors of the study hypothesized that the phenotype is due to the loss of an as yet unidentified gene or is a result of small changes in a large number of known target genes.

Endothelial homeostasis, driven by shear stress–dependent gene expression, is an important regulator of protective responses to acute and chronic inflammation [68]. Inflammation plays a central role in the development of atherosclerosis [69]. In response to inflammatory mediators, the expression of adhesion molecules such as vascular cell adhesion molecule-1 (VCAM-1) is induced on the surface of VE cells. Adhesion molecules mediate the recruitment of leukocytes to the endothelium, and their expression is a crucial step in the formation of atherosclerotic lesions. Our group demonstrated that KLF2 functions as an anti-inflammatory factor in endothelial cells and that it inhibits cytokine-mediated VCAM-1 induction [29]. NF-κB is a well-established key regulator of cytokine-induced endothelial cell activation. We have demonstrated that KLF2 inhibits NF-κB function by inhibiting recruitment of CBP/p300, a NF-κB cofactor critical in mediating NF-κB activity. Binding of KLF2 to CBP/p300 inhibits the interaction between p65 and CBP/p300. As a consequence, NF-κB is not optimally activated [29].

TNF-α signaling is a major driver of endothelial inflammation in atherosclerosis and is mediated through a NF-κB and MAPK mechanism that results in the upregulation of activation protein 1 (AP-1) [70]. Shear stress–induced KLF2 inhibits activating transcription factor 2 (ATF2), one of the heterodimeric components of AP-1 [71]. Increased levels of phosphorylated ATF2 are seen in endothelial cells overlying early atherosclerotic plaques. Knockdown studies using small interfering RNA (siRNA) against KLF2 suppressed the inhibitory effect of shear stress on ATF2. Furthermore, KLF2 excluded phosphorylated ATF2 from the nucleus [71], thereby inhibiting ATF2's ability to activate inflammatory pathways.

Transforming growth factor-β (TGF-β) has been proposed to regulate endothelial cell activation [72]. KLF2 regulates the TGF-β-mediated proinflammatory response in endothelial cells by inducing Smad7, a potent attenuator of TGF-β signaling. This results in a reduction of the phosphorylation and nuclear accumulation of Smad2 and the Smad3/4-dependent transcriptional cascade, thereby abrogating TGF-β signaling. Furthermore, it was demonstrated that

KLF2 can inhibit AP-1, an important cofactor in TGF-β-mediated inflammatory activity [73].

Consistent with these in vitro observations, hemizygous deficiency of KLF2 promotes diet-induced atherosclerosis in apolipoprotein E–deficient mice [33]. Hemizygous deficiency of KLF2 did not, however, show significant changes in endothelial mediators of inflammation, including VCAM-1, eNOS, and TM. This may be due to a compensatory effect of KLF4 (discussed below), which was increased in KLF2-deficient mice compared to control animals [33]. The mechanism of increased susceptibility to atherosclerosis may also be a result of alterations in macrophage activity; hemizygous deficiency of KLF2 enhances the expression of the key lipid-binding protein adipocyte protein 2/fatty acid–binding protein 4 and increases macrophage lipid.

Post-transcriptional regulation of KLF2 by microRNAs (miRNAs) from shear stress provides further complexity to this mechanism. MiRNAs are short, noncoding RNA sequences involved in transcriptional gene expression regulation and are increasingly being recognized as having a role in cardiovascular disease [74]. Areas of disturbed flow increase expression of miR-92A, a proinflammatory miRNA that transrepresses KLF2 and KLF4. Fang and Davies demonstrated that miR-92A has predicted binding sites to the 3'-untranslated regions of KLF2 and KLF4 and that knockdown of miR-92A rescued this KLF transrepression [75]. Furthermore, laminar shear stress decreases the expression of miR-92A, leading to an atheroprotective phenotype from increased endothelial KLF2 expression [76]. Oxidative stress also induces expression of miR-92A, leading to endothelial dysfunction from KLF2 transrepression [77]. Combining these observations was a study by Loyer et al. demonstrating that hypercholesterolemia markedly enhanced miR-92A in atherosclerotic low-density lipoprotein receptor deficient (ldlr$^{--/-}$) mice through oxidated LDL, especially in the disturbed flow atheroprone areas of the vasculature. This phenotype was rescued by inhibition of miR-92A, leading to increased KLF2 and KLF4 expression and reduced endothelial inflammation [78].

2.2.1.2.2 *Coagulation and thrombosis*

Coagulation is a highly conserved process that is critical to endothelial function. Cross talk between the coagulation and inflammatory pathways is vital to maintaining an intact and healthy endothelium as in pathological states there can be inappropriate thrombosis leading to clinical syndromes such as myocardial infarction or pulmonary embolism [79–81]. KLF2 and KLF4 regulate numerous steps that protect the healthy endothelium from pathological thrombosis. Global overexpression of KLF2 postnatal in mice produces an antithrombotic phenotype with prolonged clotting times in the carotid thrombosis assay model. Deletion of KLF2 produced an antiparallel effect with reduced clotting times [82].

Inflammatory stimuli induce the expression of tissue factor (TF) and plasminogen activator inhibitor 1 (PAI-1) by endothelial cells, resulting in a local imbalance of anticoagulant and procoagulant proteins and the development of clots. We have shown that KLF2 inhibits cytokine-mediated PAI-1 and TF induction in endothelial cells [27]. The relationship between shear stress and endothelial expression of coagulation proteins has been described for several of the factors regulated by KLF2. While a transient *increase* in TF expression is seen in cultured endothelial cells exposed to laminar shear stress [83], in vivo studies in rat carotid arteries show TF expression only in vessels subjected to mechanical stenosis [84]. Analysis of human autopsy specimens demonstrates a selective lack of expression of TF and PAI-1 in regions of the carotid artery exposed to atheroprotective flow [85]. KLF2 robustly induces TM, a potent inhibitor of blood coagulation [27].

Consistent with these observations, overexpression of KLF2 in human umbilical vein endothelial cells (HUVECs) increases blood clotting time under flow conditions. Conversely, knockdown of KLF2 by siRNA leads to a reduced time to clot formation. Our mechanistic studies demonstrate that KLF2 binds to specific sites within the TM promoter and thereby induces transcription of the TM promoter [27]. TM expression is enhanced by fluid shear stress [86, 87]. In cultured endothelial cells made deficient in KLF2 with siRNA and exposed to atheroprotective flow, upregulation of TM is significantly diminished [57]. Thrombin is a key component of the coagulation cascade and has been shown to induce inflammatory gene expression via NF-κB activation [88]. Lin and colleagues demonstrated that KLF2 inhibits thrombin-mediated expression of proinflammatory mediators and chemokines, including interleukin-6 (IL-6), IL-8, and monocyte chemoattractant protein-1 (MCP-1), in endothelial cells

[30]. Most of thrombin's activity is mediated via protease-activated receptor 1 (PAR-1), a G-coupled protein receptor. Thrombin cleaves the receptor, unmasking a tethered ligand that then activates the receptor [79, 89, 90]. KLF2 negatively regulates PAR-1 expression and inhibits thrombin-mediated NF-κB activation [30]. Interestingly, PAR-1 expression is increased at sites of vascular injury (and thus altered flow mechanics) but downregulated by laminar shear stress [91]. In addition, TNF-α is an inflammatory cytokine that promotes inflammation and coagulation through the repression of endothelial TM via an NF-κB-dependent pathway. IKKβ is an upstream regulatory kinase of NF-κB that when activated will phosphorylate a protein named inhibitor of NF-κB (IkB) that will bind to NF-κB and inhibit its signaling. Inhibition of IKKβ increases TM expression and function through the overexpression and binding of KLF2 to the TM promoter [92].

Exposure to ambient particulate matter (PM) in air pollution has long been correlated with increased cardiovascular morbidity and mortality. Urban PM arises from a combination of dust, soot, smoke, and combustion and was found to downregulate KLF2 and TM expression, leading to increased coagulation in the lungs of PM-exposed wild. The class III HDAC Sirt1 regulates lung inflammation and coagulation by inhibiting the downregulation of KLF2 and TM expression, thereby reducing lung coagulation after PM exposure [93].

Antiphospholipid syndrome is an autoimmune disease characterized by recurrent arterial and venous thrombosis in the presence of antiphospholipid antibodies (APLA). It is most frequently seen in systemic lupus erythematosus. The majority of APLA antagonize phospholipid-binding proteins, particularly against β2-glycoprotein I (β_2GPI). Anti-β_2GPI antibodies activate endothelial cells to induce inflammation through the downregulation of KLF2/KLF4. This in turn leads to upregulation of NF-κB and the expression of proinflammatory molecules. As described earlier, KLFs sequester the cotranscriptional activator CBP/p300, making it unavailable for NF-κB. Competition between KLF and NF-κB for CBP/p300 is likely the mechanism of this phenotype as restoration of KLF2 or KLF4 expression inhibits NF-κB transactivation and blocks APLA/anti-β_2GPI-mediated endothelial activation [94].

2.2.1.2.3 *Vascular development/maturation/remodeling*

Shear stress appears to control key aspects of embryonic cardiovascular development [95] as well as remodeling of developing and mature arteries [96, 97]. KLF2 knockout mice are embryonic lethal at E12.5–14.5 due to severe intraembryonic and intra-amniotic hemorrhaging [42, 43]. In this model, the formation of the vascular network appeared normal and both vasculogenesis and arteriogenesis were intact, but the vessel morphology was abnormal. KLF2$^{-/-}$ mice showed umbilical vessel and aortic defects resulting from impaired smooth muscle cell recruitment and tunica media formation, causing aortic dilation and rupture [42]. Of note, significant structural vascular defects (hemorrhage, aneurysm, or abnormal smooth muscle morphology) were not observed by Lee et al. [67] in KLF2$^{-/-}$ mice, with the exception of a few late-stage embryos. These authors explain the apparent discrepancy by interpreting the morphological defects to be a late, secondary effect of the primary physiologic event (low vascular tone). Lingrel and colleagues have subsequently demonstrated that KLF2$^{-/-}$ embryos have a normal development pattern of endothelial cells but a reduced number of smooth muscle cells in the aorta. Deficiency of KLF2 also inhibited smooth muscle cell migration in response to platelet-derived growth factor B (PDGF-B), resulting in a lack of vessel maturation and stability [98]. Furthermore, shear stress waveforms designed to mimic those seen in coronary collateral vessels induce KLF2 as well as genes important for endothelium-smooth muscle interactions [99].

It turns out that a significant portion of the endothelial transcriptome is regulated by KLF2 and KLF4 and that expression of at least one endothelial KLF is necessary for survival. As described above, systemic deletion of KLF2 and/or KLF4 results in death during embryonic development or shortly after birth. Endothelial-specific deletion of KLF2 and KLF4 results in acute vascular dysfunction with a loss of vascular integrity and systemic consumptive coagulopathy, leading to death from myocardial infarction, heart failure, and stroke. Single expression of either endothelial KLF2 or KLF4 is sufficient for survival, suggesting the presence of some redundancy between the two KLFs. Gene expression analysis of the endothelial-specific KLF2/KLF4 double-deletion mouse shows profound alterations of the endothelial transcriptome, with changes in over 6000 genes, compared to the Cre control. In contrast, single deletion of either KLF2 (45 genes) or KLF4 (30 genes) resulted in only a small number of differentially expressed genes [100]. Although single expression of either KLF2 or KLF4 allows for survival into adulthood, there is still significant endothelial dysfunction, with increased susceptibility to stressors, such as ischemia, chronic inflammation, atherosclerosis, and injury-induced neointimal formation, with the loss of either KLF [101–103]. Together these data highlight the critical role of KLF2 and KLF4 in endothelial development and function.

2.2.1.2.4 *Angiogenesis*

Angiogenesis is a complex process that involves multiple gene products expressed by different cell types and recapitulates many of the molecular events that occur during vascular development. Imbalance of this process results in malignant and ischemic disorders [104]. The effect of shear stress on angiogenesis is not well described; most of the expected effects are extrapolated from the pattern of growth factors, such as VEGF, induced by in vitro shear stress experiments. Studies from our laboratory demonstrate that KLF2 potently inhibits VEGF-mediated angiogenesis [31]. In the nude mouse ear model of angiogenesis, KLF2 potently inhibits permeability, tissue edema, and angiogenesis. From a mechanistic standpoint, KLF2 inhibits cell proliferation, VEGFR2 expression, and VEGF-induced endothelial activation (characterized by reduction of calcium influx and suppression of VCAM-1, TF, and cyclooxygenase 2) [31]. KLF2 has been shown to regulate VEGFR2 (also referred as to FMS-like tyrosine kinase 1 [Flt1] or kinase insert domain receptor [KDR]) by cooperating with Ets during vascular development in *Xenopus* embryos [105].

Ang-1 and -2 are Tie ligands that have critical roles in regulating angiogenesis [106, 107]. Ang-1-mediated Tie2 activation is required to maintain the quiescent resting state of the endothelium. Agonistic Ang-1 functions are antagonized by Ang-2, which is believed to inhibit Ang-1/Tie2 signaling. Ang-2 destabilizes the quiescent endothelium and primes it to respond to exogenous stimuli, thereby facilitating the activities of inflammatory and angiogenic cytokines [106]. KLF2 has been shown to inhibit cytokine-mediated Ang-2 expression in endothelial cells [57]. Furthermore, KLF2 has been shown to induce Tie2 expression [57]. A fundamental aspect of sprouting angiogenesis is the migration of endothelial cells. Studies from the Horrevoetts laboratory showed that KLF2 inhibits endothelial cell migration by upregulating semaphorin-3F, a potent antimigratory factor [44]. Another group demonstrated that KLF2 also downregulates proliferation, migration, and tube formation through the inhibition of ERK1/2 phosphorylation [108].

As described above, KLF2 has been implicated as an antiangiogenic factor. However, the role of KLF2 in hypoxia-mediated angiogenesis has not been elucidated. Our laboratory identified KLF2 as a negative regulator of hypoxia-mediated angiogenesis. KLF2 protein expression is acutely upregulated by hypoxia. Adenoviral overexpression of KLF2 inhibited matrigel tube formation whereas primary micro VE cells from $KLF2^{+/-}$ mice showed enhanced tube formation in response to hypoxia. Hypoxia-inducible factor-1α (HIF-1α) is a key regulator of angiogenesis under hypoxic conditions (e.g., ischemic heart disease and cancer). We showed that KLF2 inhibits HIF-1α expression and function in the endothelial cells [109]. Interestingly, KLF2 limits the accumulation of HIF-1α protein in response to hypoxia. Loss of function studies using siRNA-mediated KLF2 knockdown and mouse $KLF2^{-/-}$ embryonic fibroblasts showed accelerated HIF-1α accumulation in response to hypoxia. We found that KLF2 promotes HIF-1α degradation in a von Hippel Lindau protein–independent, p53-independent, but proteasome-dependent pathway [109]. The mechanism of increased HIF1α degradation may be explained by the ability of KLF2 to disrupt the interaction between HIF-1α and its chaperone Hsp90, suggesting that KLF2 promotes degradation of HIF-1α by affecting its folding and maturation [109]. These observations indicate that KLF2 negatively regulates hypoxia-mediated angiogenesis by inhibiting HIF-1α expression and function.

A novel concept has emerged demonstrating the role of laminar shear stress in regulating endothelial metabolism through KLF2 and its effects of angiogenesis. Endothelial cells rely heavily on glycolysis for ATP production during angiogenesis [110]. It turns out that KLF2 is a potent inhibitor of glycolysis through the inhibition of several key glycolytic enzymes, including 6-phosphofructo-2-kinase/fructose-2,6-biphosphatase-3, phosphofructokinase-1, and hexokinase 2.

Mechanotransduction signaling through shear stress is a driver of this KLF2-induced repression of glycolysis and angiogenesis [108].

2.2.1.2.5 *Vascular stress/injury*

While oscillatory or absent shear stress in areas of disturbed blood flow enables oxidative stress to take full effect, laminar shear stress results in the expression of several antioxidative factors [111]. Several studies have identified novel factors that are induced by laminar shear stress in a KLF2-dependent manner. Laminar shear stress induces expression and nuclear accumulation of nuclear factor erythroid 2–related factor 2 (Nrf2), an antioxidant transcription factor that activates antioxidant response element–dependent transcriptional programs [112]. Interestingly, the expression of Nrf2 and its target genes, such as NAD(P)H dehydrogenase quinone 1 (*NQO1*) and heme oxygenase (*HO-1*), was upregulated by laminar shear stress and KLF2. Knockdown of KLF2 resulted in the loss of induction of NQO1 but not Nrf2 by shear stress. KLF2 does, however, enhance the nuclear localization and activation of Nrf2 and promotes its antioxidant activity [112]. Together KLF2 and Nrf2 govern approximately 70% of shear stress–induced genes [73, 112].

Mason and colleagues have identified CD59 as a novel factor that is induced by laminar shear stress in a KLF2-dependent fashion [113]. Complement activation occurs in vascular injury and is thought to have the capability to induce a proinflammatory response in endothelial cells. CD59 blocks the terminal pathway of complement activation. Functionally, the increase of CD59 by laminar shear stress reduces C9 deposition and complement lysis. CD59 induction is dependent upon the magnitude of shear force. This characteristic is manifested in its differential expression pattern in murine aorta [113]. Mechanistic analysis using inhibitors of signaling pathways reveals that CD59 induction is independent of PI3K, ERK1/2, and NO. Knockdown studies by use of siRNA have demonstrated that the induction is ERK5 and KLF2 dependent [113]. These observations indicate that CD59 is a KLF2-dpendent antiatherogenic factor that is induced by laminar shear stress in endothelial cells.

Barrier integrity is another key function of a healthy endothelium regulated by KLF2. The blood–brain barrier is an important structure comprising endothelial cells and controls the trafficking of solutes from the blood into the cerebrospinal fluid. Using a transient ischemic model for stroke via middle cerebral artery occlusion, we found that transgenic KLF2–overexpressing mice were protected from ischemic stroke and had preserved the blood–brain barrier function [114]. Mechanistically this occurs through KLF2-mediated upregulation of the junction protein occludin and phosphorylation of the myosin light chain [103, 114].

2.2.1.2.6 *KLF2, shear stress, and statins*

Statins (3-hydroxy-3-methylglutaryl coenzyme A inhibitors) are widely used in the treatment of dyslipidemia [115]. As demonstrated by basic and clinical studies, statins also have pleiotropic effects apart from lowering cholesterol, including anti-inflammatory and antithrombotic effects. As demonstrated in the JUPITER trial, treatment of patients with evidence of clinical inflammation based on elevated high-sensitivity C-reactive protein levels but who were otherwise healthy and without clinical hyperlipidemia using rosuvastatin led to a reduction of cardiovascular events [116]. The effect of statins on endothelial function overlaps with that of KLF2. This fact led us to speculate whether there might be the novel link between KLF2 and statins. Indeed, we and others have demonstrated that multiple statins robustly induce KLF2 in endothelial cells [117, 118]. Importantly, knockdown of KLF2 inhibits statin-mediated eNOS and TM induction, indicating that KLF2 mediates statin effects [117]. Both MEF2 and Rho activity is critical for regulating KLF2 expression by statins. MEF2 directly transactivates the KLF2 promoter. Constitutive Rho activity downregulates KLF2 by geranylgeranyl pyrophosphate–dependent mechanisms [117]. By inhibiting the conversion of 3-hydroxy-3-methyl-glutaryl-coenzyme A (HMG-CoA) to mevalonate, statins lead to the depletion of geranylgeranyl pyrophosphate.

Endothelial sensitivity to statins depends on shear stress, and this effect is mediated by KLF2 [119]. HO-1 is a rate-limiting factor in the catabolism of heme into biliverdin and functions as an antiatherogenic factor [120]. HO-1 has been shown to be upregulated by laminar shear stress [111] and statin [121] in endothelial cells. These investigators demonstrated that preconditioning endothelial cells with laminar shear stress reduces the concentration of statin required to enhance HO-1-mediated cytoprotection against oxidant-induced injury [119]. Knockdown of KLF2 did not alter HO-1 induction by shear stress alone; however, it resulted in inhibition of HO-1 induction by the combination of shear stress and statins [119]. These observations indicate that KLF2 mediates a synergistic vasoprotective effect of shear stress and statins. Taken together, these observations indicate that KLF2 is a mediator of pleiotropic effects of statins and has significant implications in the treatment of cardiovascular disease.

2.2.2 Krüppel-Like Factor 4

KLF4 is highly expressed in terminally differentiated epithelial cells and was first cloned as gut-enriched KLF by Yang and colleagues in 1996 [122]. Gastric epithelial deletion of KLF4 causes precancerous changes, including altered proliferation and differentiation, suggesting the critical role of KLF4 in normal gastric epithelial homeostasis [123]. Furthermore, KLF4 has been implicated as a crucial factor in the development and maintenance of mouse cornea [124, 125]. KLF4 plays an important role in skin barrier function [126]. Segre and colleagues showed that KLF4$^{-/-}$ mice die shortly after birth; death was postulated to be due to dehydration resulting from loss of skin barrier function [126].

KLF4 is expressed in many cell types besides epithelial cells and has been shown to have significant roles in endothelial cell biology [127], vascular smooth muscle cell biology [128–131], macrophage biology [132, 133], and tumor biology [134–136]. In 2006, Yamanaka and colleagues identified KLF4 as a crucial factor in generating inducible pluripotent stem cells [137]. Studies from this group and others demonstrate that introduction of Oct3/4, Sox2, c-Myc, and KLF4 by retrovirus or plasmid can induce reprogram mouse fibroblasts into pluripotent stem cells [138–140].

The first description of KLF4 in endothelial cells was provided by Yet and colleagues in 1998 [141]. A subsequent study by McCormick and colleagues identified *KLF4* as a shear stress–inducible gene by DNA microarray analysis of HUVECs [142]. Hamik and colleagues first revealed in vivo expression of endothelial cells and demonstrated that KLF4 negatively regulates endothelial inflammation. The induction of KLF4 by shear stress was dependent upon the magnitude of shear force; arterial stress was more potent than venous shear stress [127].

In the endothelium, KLF4 functions very similarly to KLF2 as they share many of the same transcriptional targets, are strongly activated by laminar shear stress, and are repressed by disturbed flow [127, 143]. Just like KLF2, KLF4 interacts with p65 in the NF-κB signaling pathway to inhibit inflammatory signaling and expression of VCAM-1 [101, 102]. Overexpression of KLF4 induced transactivation of atheroprotective genes, such as *eNOS* and *TM* [127, 144]. Furthermore, KLF4 reduced cytokine-mediated VCAM-1, PAI-1, and TF induction in endothelial cells. Consistent with these observations, KLF4 inhibited cytokine-mediated monocyte adhesion on endothelial cells. Using the prolonged carotid artery thrombosis model to study atherothrombosis, we showed that the deletion of endothelial KLF4 created a prothrombotic phenotype that was reversed in the overexpression model [101]. KLF4 expression is also upregulated by statins and confers a vasoprotective phenotype [144]. MiRNAs are also involved in KLF4 signaling, in particular miR-92a, which was described earlier as being a proatherogenic with preferential expression in areas of disturbed flow and having a transrepressive effect on KLF2/4 [78, 144]. KLF4 is also uniquely regulated by miRNA-103 in a transrepressive mechanism to increase NF-κB signaling and C-X-C-motif chemokine 1–mediated macrophage recruitment to atherosclerotic lesions in the murine.

2.3 Future Directions

KLFs are increasingly being recognized as master regulators of the endothelial phenotype. Shear stress is an important regulator of normal endothelial function conferring numerous vasoprotective effects, including vasodilatory, anti-inflammatory, and antithrombotic effects, that all coalesce through KLF signal transduction. All key functions of endothelial physiology are regulated by the KLFs, and their downregulation is associated with pathology that manifests clinically as cardiovascular disease (described in Fig. 2.2). Redundancy of the endothelial transcriptome is vital for endothelial development and function. KLF2 and KLF4 overlap in the numerous functions that are critical to development and endothelial physiology. The expression of over 6000 genes was profoundly altered in the KLF2/4 double-deletion mouse, while single deletion of either KLF2 or KLF4 resulted in only a small number of differentially expressed genes (less than 50) [100]. Loss of KLF2 leads to a compensatory upregulation of KLF4 expression as the heterozygous KLF2 apolipoprotein E knockout ($ApoE^{-/-}$) null mouse lacks proinflammatory expression, possibly due to a compensatory increase in KLF4 expression [33]. Single deletion of either KLF2 and KLF4 still leads to significant endothelial dysfunction and increased susceptibility to stressors in adulthood [101–103].

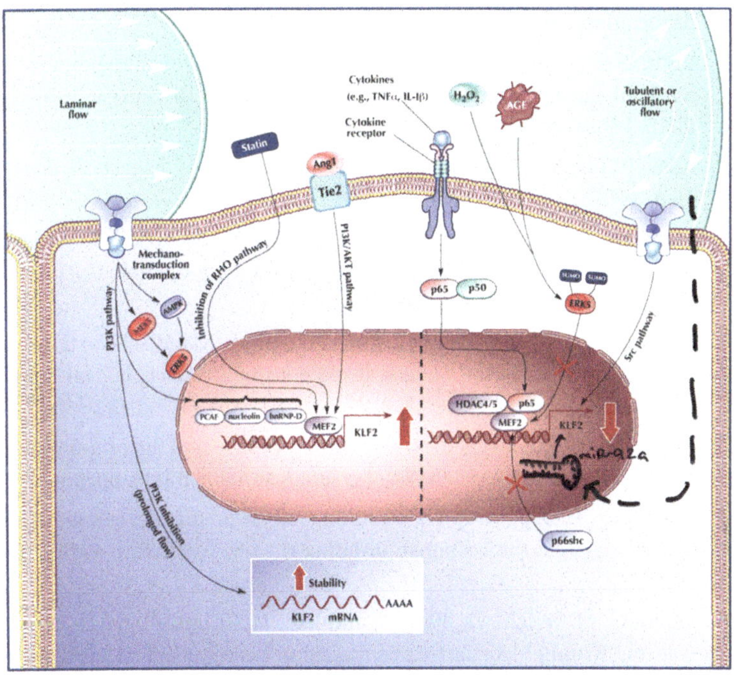

Figure 2.2 Schematic overview of the role of endothelial KLFs in shear stress–induced vasoprotection. See text for details.

Vascular aging and the role of autophagy in the endothelium have garnered increasing interest recently. Autophagy is a conserved physiological process through which the cell is able to recycle damaged proteins and cell organelles. We recently demonstrated the role of KLF4 in a protective augmenting autophagy and improving vascular function in aged mice [145]. We hypothesize that perhaps some of the redundancies between KLF2 and KLF4 may change with aging and help define the distinct differences between these endothelial KLFs. Further investigations are warranted to elucidate these mechanisms and identify potential pharmacological targets for therapy.

References

1. Mozaffarian, D., Benjamin, E. J., Go, A. S., et al. (2015). Heart disease and stroke statistics--2015 update: a report from the American Heart Association. *Circulation*, **131**(4), pp. e29–322.
2. Roth, G. A., Johnson, C., Abajobir, A., et al. (2017). Global, regional, and national burden of cardiovascular diseases for 10 causes, 1990 to 2015. *J Am Coll Cardiol*, **70**(1), pp. 1–25.
3. Ross, R. (1999). Atherosclerosis--an inflammatory disease. *N Engl J Med*, **340**(2), pp. 115–126.
4. Monahan-Earley, R., Dvorak, A. M. and Aird, W. C. (2013). Evolutionary origins of the blood vascular system and endothelium. *J Thromb Haemost*, **11**(Suppl 1), pp. 46–66.
5. Baeyens, N., Bandyopadhyay, C., Coon, B. G., Yun, S. and Schwartz, M. A. (2016). Endothelial fluid shear stress sensing in vascular health and disease. *J Clin Invest*, **126**(3), pp. 821–828.
6. Glagov, S. (1973). Mechanical stresses on vessels and the non-uniform distribution of atherosclerosis. *Med Clin North Am*, **57**(1), pp. 63–77.
7. Weinberg, P. D. (2004). Rate-limiting steps in the development of atherosclerosis: the response-to-influx theory. *J Vasc Res*, **41**(1), pp. 1–17.
8. Feinberg, M. W., Lin, Z., Fisch, S. and Jain, M. K. (2004). An emerging role for Kruppel-like factors in vascular biology. *Trends Cardiovasc Med*, **14**(6), pp. 241–246.
9. Atkins, G. B. and Jain, M. K. (2007). Role of Kruppel-like transcription factors in endothelial biology. *Circ Res*, **100**(12), pp. 1686–1695.
10. Haldar, S. M., Ibrahim, O. A. and Jain, M. K. (2007). Kruppel-like factors (KLFs) in muscle biology. *J Mol Cell Cardiol*, **43**(1), pp. 1–10.
11. Hamik, A., Wang, B. and Jain, M. K. (2006). Transcriptional regulators of angiogenesis. *Arterioscler Thromb Vasc Biol*, **26**(9), pp. 1936–1947.
12. Jackle, H., Rosenberg, U. B., Preiss, A., et al. (1985). Molecular analysis of Kruppel, a segmentation gene of Drosophila melanogaster. *Cold Spring Harb Symp Quant Biol*, **50**, pp. 465–473.

13. Preiss, A., Rosenberg, U. B., Kienlin, A., Seifert, E. and Jackle, H. (1985). Molecular genetics of Kruppel, a gene required for segmentation of the Drosophila embryo. *Nature,* **313**(5997), pp. 27–32.
14. Rosenberg, U. B., Preiss, A., Seifert, E., Jackle, H. and Knipple, D. C. (1985). Production of phenocopies by Kruppel antisense RNA injection into Drosophila embryos. *Nature,* **313**(6004), pp. 703–706.
15. Miller, I. J. and Bieker, J. J. (1993). A novel, erythroid cell-specific murine transcription factor that binds to the CACCC element and is related to the Kruppel family of nuclear proteins. *Mol Cell Biol,* **13**(5), pp. 2776–2786.
16. Black, A. R., Black, J. D. and Azizkhan-Clifford, J. (2001). Sp1 and Kruppel-like factor family of transcription factors in cell growth regulation and cancer. *J Cell Physiol,* **188**(2), pp. 143–160.
17. van Vliet, J., Crofts, L. A., Quinlan, K. G., Czolij, R., Perkins, A. C. and Crossley, M. (2006). Human KLF17 is a new member of the Sp/KLF family of transcription factors. *Genomics,* **87**(4), pp. 474–482.
18. Chang, E., Nayak, L. and Jain, M. K. (2017). Kruppel-like factors in endothelial cell biology. *Curr Opin Hematol,* **24**(3), pp. 224–229.
19. McConnell, B. B. and Yang, V. W. (2010). Mammalian Kruppel-like factors in health and diseases. *Physiol Rev,* **90**(4), pp. 1337–1381.
20. Pei, J. and Grishin, N. V. (2013). A new family of predicted Kruppel-like factor genes and pseudogenes in placental mammals. *PLoS One,* **8**(11), pp. e81109.
21. Dang, D. T., Zhao, W., Mahatan, C. S., Geiman, D. E. and Yang, V. W. (2002). Opposing effects of Kruppel-like factor 4 (gut-enriched Kruppel-like factor) and Kruppel-like factor 5 (intestinal-enriched Kruppel-like factor) on the promoter of the Kruppel-like factor 4 gene. *Nucleic Acids Res,* **30**(13), pp. 2736–2741.
22. Zhang, W., Shields, J. M., Sogawa, K., Fujii-Kuriyama, Y. and Yang, V. W. (1998). The gut-enriched Kruppel-like factor suppresses the activity of the CYP1A1 promoter in an Sp1-dependent fashion. *J Biol Chem,* **273**(28), pp. 17917–17925.
23. Wani, M. A., Wert, S. E. and Lingrel, J. B. (1999). Lung Kruppel-like factor, a zinc finger transcription factor, is essential for normal lung development. *J Biol Chem,* **274**(30), pp. 21180–21185.
24. Conkright, M. D., Wani, M. A. and Lingrel, J. B. (2001). Lung Kruppel-like factor contains an autoinhibitory domain that regulates its transcriptional activation by binding WWP1, an E3 ubiquitin ligase. *J Biol Chem,* **276**(31), pp. 29299–29306.

25. Zhang, X., Srinivasan, S. V. and Lingrel, J. B. (2004). WWP1-dependent ubiquitination and degradation of the lung Kruppel-like factor, KLF2. *Biochem Biophys Res Commun*, **316**(1), pp. 139–148.

26. Anderson, K. P., Kern, C. B., Crable, S. C. and Lingrel, J. B. (1995). Isolation of a gene encoding a functional zinc finger protein homologous to erythroid Kruppel-like factor: identification of a new multigene family. *Mol Cell Biol*, **15**(11), pp. 5957–5965.

27. Lin, Z., Kumar, A., SenBanerjee, S., et al. (2005). Kruppel-like factor 2 (KLF2) regulates endothelial thrombotic function. *Circ Res*, **96**(5), pp. e48–57.

28. Kumar, A., Lin, Z., SenBanerjee, S. and Jain, M. K. (2005). Tumor necrosis factor alpha-mediated reduction of KLF2 is due to inhibition of MEF2 by NF-kappaB and histone deacetylases. *Mol Cell Biol*, **25**(14), pp. 5893–5903.

29. SenBanerjee, S., Lin, Z., Atkins, G. B., et al. (2004). KLF2 Is a novel transcriptional regulator of endothelial proinflammatory activation. *J Exp Med*, **199**(10), pp. 1305–1315.

30. Lin, Z., Hamik, A., Jain, R., Kumar, A. and Jain, M. K. (2006). Kruppel-like factor 2 inhibits protease activated receptor-1 expression and thrombin-mediated endothelial activation. *Arterioscler Thromb Vasc Biol*, **26**(5), pp. 1185–1189.

31. Bhattacharya, R., Senbanerjee, S., Lin, Z., et al. (2005). Inhibition of vascular permeability factor/vascular endothelial growth factor-mediated angiogenesis by the Kruppel-like factor KLF2. *J Biol Chem*, **280**(32), pp. 28848–28851.

32. Das, H., Kumar, A., Lin, Z., et al. (2006). Kruppel-like factor 2 (KLF2) regulates proinflammatory activation of monocytes. *Proc Natl Acad Sci U S A*, **103**(17), pp. 6653–6658.

33. Atkins, G. B., Wang, Y., Mahabeleshwar, G. H., et al. (2008). Hemizygous deficiency of Kruppel-like factor 2 augments experimental atherosclerosis. *Circ Res*, **103**(7), pp. 690–693.

34. Kuo, C. T., Veselits, M. L. and Leiden, J. M. (1997). LKLF: a transcriptional regulator of single-positive T cell quiescence and survival. *Science*, **277**(5334), pp. 1986–1990.

35. Wu, J. and Lingrel, J. B. (2005). Kruppel-like factor 2, a novel immediate-early transcriptional factor, regulates IL-2 expression in T lymphocyte activation. *J Immunol*, **175**(5), pp. 3060–3066.

36. Carlson, C. M., Endrizzi, B. T., Wu, J., et al. (2006). Kruppel-like factor 2 regulates thymocyte and T-cell migration. *Nature*, **442**(7100), pp. 299–302.

37. Sebzda, E., Zou, Z., Lee, J. S., Wang, T. and Kahn, M. L. (2008). Transcription factor KLF2 regulates the migration of naive T cells by restricting chemokine receptor expression patterns. *Nat Immunol*, **9**(3), pp. 292–300.

38. Basu, P., Morris, P. E., Haar, J. L., et al. (2005). KLF2 is essential for primitive erythropoiesis and regulates the human and murine embryonic beta-like globin genes in vivo. *Blood*, **106**(7), pp. 2566–2571.

39. Banerjee, S. S., Feinberg, M. W., Watanabe, M., et al. (2003). The Kruppel-like factor KLF2 inhibits peroxisome proliferator-activated receptor-gamma expression and adipogenesis. *J Biol Chem*, **278**(4), pp. 2581–2584.

40. Wu, J., Srinivasan, S. V., Neumann, J. C. and Lingrel, J. B. (2005). The KLF2 transcription factor does not affect the formation of preadipocytes but inhibits their differentiation into adipocytes. *Biochemistry*, **44**(33), pp. 11098–11105.

41. Jiang, J., Chan, Y. S., Loh, Y. H., et al. (2008). A core KLF circuitry regulates self-renewal of embryonic stem cells. *Nat Cell Biol*, **10**(3), pp. 353–360.

42. Kuo, C. T., Veselits, M. L., Barton, K. P., Lu, M. M., Clendenin, C. and Leiden, J. M. (1997). The LKLF transcription factor is required for normal tunica media formation and blood vessel stabilization during murine embryogenesis. *Genes Dev*, **11**(22), pp. 2996–3006.

43. Wani, M. A., Means, R. T., Jr. and Lingrel, J. B. (1998). Loss of LKLF function results in embryonic lethality in mice. *Transgenic Res*, **7**(4), pp. 229–238.

44. Dekker, R. J., Boon, R. A., Rondaij, M. G., et al. (2006). KLF2 provokes a gene expression pattern that establishes functional quiescent differentiation of the endothelium. *Blood*, **107**(11), pp. 4354–4363.

45. Dekker, R. J., van Soest, S., Fontijn, R. D., et al. (2002). Prolonged fluid shear stress induces a distinct set of endothelial cell genes, most specifically lung Kruppel-like factor (KLF2). *Blood*, **100**(5), pp. 1689–1698.

46. Dekker, R. J., van Thienen, J. V., Rohlena, J., et al. (2005). Endothelial KLF2 links local arterial shear stress levels to the expression of vascular tone-regulating genes. *Am J Pathol*, **167**(2), pp. 609–618.

47. Groenendijk, B. C., Hierck, B. P., Vrolijk, J., et al. (2005). Changes in shear stress-related gene expression after experimentally altered venous return in the chicken embryo. *Circ Res*, **96**(12), pp. 1291–1298.

48. Groenendijk, B. C., van der Heiden, K., Hierck, B. P. and Poelmann, R. E. (2007). The role of shear stress on ET-1, KLF2, and NOS-3 expression

in the developing cardiovascular system of chicken embryos in a venous ligation model. *Physiology (Bethesda)*, **22**, pp. 380–389.

49. van Thienen, J. V., Fledderus, J. O., Dekker, R. J., et al. (2006). Shear stress sustains atheroprotective endothelial KLF2 expression more potently than statins through mRNA stabilization. *Cardiovasc Res*, **72**(2), pp. 231–240.

50. Boon, R. A., Leyen, T. A., Fontijn, R. D., et al. (2010). KLF2-induced actin shear fibers control both alignment to flow and JNK signaling in vascular endothelium. *Blood*, **115**(12), pp. 2533–2542.

51. Huddleson, J. P., Ahmad, N., Srinivasan, S. and Lingrel, J. B. (2005). Induction of KLF2 by fluid shear stress requires a novel promoter element activated by a phosphatidylinositol 3-kinase-dependent chromatin-remodeling pathway. *J Biol Chem*, **280**(24), pp. 23371–23379.

52. Huddleson, J. P., Srinivasan, S., Ahmad, N. and Lingrel, J. B. (2004). Fluid shear stress induces endothelial KLF2 gene expression through a defined promoter region. *Biol Chem*, **385**(8), pp. 723–729.

53. Storck, S., Shukla, M., Dimitrov, S. and Bouvet, P. (2007). Functions of the histone chaperone nucleolin in diseases. *Subcell Biochem*, **41**, 125–144.

54. Huddleson, J. P., Ahmad, N. and Lingrel, J. B. (2006). Up-regulation of the KLF2 transcription factor by fluid shear stress requires nucleolin. *J Biol Chem*, **281**(22), pp. 15121–15128.

55. Wang, N., Miao, H., Li, Y. S., et al. (2006). Shear stress regulation of Kruppel-like factor 2 expression is flow pattern-specific. *Biochem Biophys Res Commun*, **341**(4), pp. 1244–1251.

56. Sohn, S. J., Li, D., Lee, L. K. and Winoto, A. (2005). Transcriptional regulation of tissue-specific genes by the ERK5 mitogen-activated protein kinase. *Mol Cell Biol*, **25**(19), pp. 8553–8566.

57. Parmar, K. M., Larman, H. B., Dai, G., et al. (2006). Integration of flow-dependent endothelial phenotypes by Kruppel-like factor 2. *J Clin Invest*, **116**(1), pp. 49–58.

58. Young, A., Wu, W., Sun, W., et al. (2009). Flow activation of AMP-activated protein kinase in vascular endothelium leads to Kruppel-like factor 2 expression. *Arterioscler Thromb Vasc Biol*, **29**(11), pp. 1902–1908.

59. Woo, C. H., Shishido, T., McClain, C., et al. (2008). Extracellular signal-regulated kinase 5 SUMOylation antagonizes shear stress-induced antiinflammatory response and endothelial nitric oxide synthase expression in endothelial cells. *Circ Res*, **102**(5), pp. 538–545.

60. Sako, K., Fukuhara, S., Minami, T., et al. (2009). Angiopoietin-1 induces Kruppel-like factor 2 expression through a phosphoinositide 3-kinase/AKT-dependent activation of myocyte enhancer factor 2. *J Biol Chem*, **284**(9), pp. 5592–5601.

61. Kumar, A., Hoffman, T. A., Dericco, J., Naqvi, A., Jain, M. K. and Irani, K. (2009). Transcriptional repression of Kruppel like factor-2 by the adaptor protein p66shc. *FASEB J*, **23**(12), pp. 4344–4352.

62. Kwon, I. S., Wang, W., Xu, S. and Jin, Z. G. (2014). Histone deacetylase 5 interacts with Kruppel-like factor 2 and inhibits its transcriptional activity in endothelium. *Cardiovasc Res*, **104**(1), pp. 127–137.

63. Wang, W., Ha, C. H., Jhun, B. S., Wong, C., Jain, M. K. and Jin, Z. G. (2010). Fluid shear stress stimulates phosphorylation-dependent nuclear export of HDAC5 and mediates expression of KLF2 and eNOS. *Blood*, **115**(14), pp. 2971–2979.

64. Lee, D. Y., Lee, C. I., Lin, T. E., et al. (2012). Role of histone deacetylases in transcription factor regulation and cell cycle modulation in endothelial cells in response to disturbed flow. *Proc Natl Acad Sci U S A*, **109**(6), pp. 1967–1972.

65. Goodwin, B. L., Solomonson, L. P. and Eichler, D. C. (2004). Argininosuccinate synthase expression is required to maintain nitric oxide production and cell viability in aortic endothelial cells. *J Biol Chem*, **279**(18), pp. 18353–18360.

66. Razani, B., Engelman, J. A., Wang, X. B., et al. (2001). Caveolin-1 null mice are viable but show evidence of hyperproliferative and vascular abnormalities. *J Biol Chem*, **276**(41), pp. 38121–38138.

67. Lee, J. S., Yu, Q., Shin, J. T., et al. (2006). Klf2 is an essential regulator of vascular hemodynamic forces in vivo. *Dev Cell*, **11**(6), pp. 845–857.

68. Hwa, C., Sebastian, A. and Aird, W. C. (2005). Endothelial biomedicine: its status as an interdisciplinary field, its progress as a basic science, and its translational bench-to-bedside gap. *Endothelium*, **12**(3), pp. 139–151.

69. Ross, R. (1999). Atherosclerosis is an inflammatory disease. *Am Heart J*, **138**(5 Pt 2), pp. S419–420.

70. Horrevoets, A. J., Fontijn, R. D., van Zonneveld, A. J., de Vries, C. J., ten Cate, J. W. and Pannekoek, H. (1999). Vascular endothelial genes that are responsive to tumor necrosis factor-alpha in vitro are expressed in atherosclerotic lesions, including inhibitor of apoptosis protein-1, stannin, and two novel genes. *Blood*, **93**(10), pp. 3418–3431.

71. Fledderus, J. O., van Thienen, J. V., Boon, R. A., et al. (2007). Prolonged shear stress and KLF2 suppress constitutive proinflammatory

transcription through inhibition of ATF2. *Blood*, **109**(10), pp. 4249–4257.

72. Goumans, M. J., Liu, Z. and ten Dijke, P. (2009). TGF-beta signaling in vascular biology and dysfunction. *Cell Res*, **19**(1), pp. 116–127.
73. Boon, R. A., Fledderus, J. O., Volger, O. L., et al. (2007). KLF2 suppresses TGF-beta signaling in endothelium through induction of Smad7 and inhibition of AP-1. *Arterioscler Thromb Vasc Biol*, **27**(3), pp. 532–539.
74. Schober, A., Nazari-Jahantigh, M. and Weber, C. (2015). MicroRNA-mediated mechanisms of the cellular stress response in atherosclerosis. *Nat Rev Cardiol*, **12**(6), pp. 361–374.
75. Fang, Y. and Davies, P. F. (2012). Site-specific microRNA-92a regulation of Kruppel-like factors 4 and 2 in atherosusceptible endothelium. *Arterioscler Thromb Vasc Biol*, **32**(4), pp. 979–987.
76. Wu, W., Xiao, H., Laguna-Fernandez, A., et al. (2011). Flow-dependent regulation of Kruppel-like factor 2 is mediated by microRNA-92a. *Circulation*, **124**(5), pp. 633–641.
77. Chen, Z., Wen, L., Martin, M., et al. (2015). Oxidative stress activates endothelial innate immunity via sterol regulatory element binding protein 2 (SREBP2) transactivation of microRNA-92a. *Circulation*, **131**(9), pp. 805–814.
78. Loyer, X., Potteaux, S., Vion, A. C., et al. (2014). Inhibition of microRNA-92a prevents endothelial dysfunction and atherosclerosis in mice. *Circ Res*, **114**(3), pp. 434–443.
79. Coughlin, S. R. (2000). Thrombin signalling and protease-activated receptors. *Nature*, **407**(6801), pp. 258–264.
80. Minami, T., Sugiyama, A., Wu, S. Q., Abid, R., Kodama, T. and Aird, W. C. (2004). Thrombin and phenotypic modulation of the endothelium. *Arterioscler Thromb Vasc Biol*, **24**(1), pp. 41–53.
81. Jialal, I., Devaraj, S. and Venugopal, S. K. (2004). C-reactive protein: risk marker or mediator in atherothrombosis? *Hypertension*, **44**(1), pp. 6–11.
82. Nayak, L., Shi, H., Atkins, G. B., Lin, Z., Schmaier, A. H. and Jain, M. K. (2014). The thromboprotective effect of bortezomib is dependent on the transcription factor Kruppel-like factor 2 (KLF2). *Blood*, **123**(24), pp. 3828–3831.
83. Lin, M. C., Almus-Jacobs, F., Chen, H. H., et al. (1997). Shear stress induction of the tissue factor gene. *J Clin Invest*, **99**(4), pp. 737–744.

84. Houston, P., Dickson, M. C., Ludbrook, V., et al. (1999). Fluid shear stress induction of the tissue factor promoter in vitro and in vivo is mediated by Egr-1. *Arterioscler Thromb Vasc Biol*, **19**(2), pp. 281–289.

85. Tohgi, H., Utsugisawa, K., Yoshimura, M., Nagane, Y. and Ukitsu, M. (1999). Local variation in expression of pro- and antithrombotic factors in vascular endothelium of human autopsy brain. *Acta Neuropathol*, **98**(2), pp. 111–118.

86. Malek, A. M., Jackman, R., Rosenberg, R. D. and Izumo, S. (1994). Endothelial expression of thrombomodulin is reversibly regulated by fluid shear stress. *Circ Res*, **74**(5), pp. 852–860.

87. Takada, Y., Shinkai, F., Kondo, S., et al. (1994). Fluid shear stress increases the expression of thrombomodulin by cultured human endothelial cells. *Biochem Biophys Res Commun*, **205**(2), pp. 1345–1352.

88. Minami, T. and Aird, W. C. (2001). Thrombin stimulation of the vascular cell adhesion molecule-1 promoter in endothelial cells is mediated by tandem nuclear factor-kappa B and GATA motifs. *J Biol Chem*, **276**(50), pp. 47632–47641.

89. Vu, T. K., Hung, D. T., Wheaton, V. I. and Coughlin, S. R. (1991). Molecular cloning of a functional thrombin receptor reveals a novel proteolytic mechanism of receptor activation. *Cell*, **64**(6), pp. 1057–1068.

90. Gerszten, R. E., Chen, J., Ishii, M., et al. (1994). Specificity of the thrombin receptor for agonist peptide is defined by its extracellular surface. *Nature*, **368**(6472), pp. 648–651.

91. Nguyen, K. T., Eskin, S. G., Patterson, C., Runge, M. S. and McIntire, L. V. (2001). Shear stress reduces protease activated receptor-1 expression in human endothelial cells. *Ann Biomed Eng*, **29**(2), pp. 145–152.

92. Pathak, R., Shao, L., Chafekar, S. M., et al. (2014). IKKbeta regulates endothelial thrombomodulin in a Klf2-dependent manner. *J Thromb Haemost*, **12**(9), pp. 1533–1544.

93. Wu, Z., Liu, M. C., Liang, M. and Fu, J. (2012). Sirt1 protects against thrombomodulin down-regulation and lung coagulation after particulate matter exposure. *Blood*, **119**(10), pp. 2422–2429.

94. Allen, K. L., Hamik, A., Jain, M. K. and McCrae, K. R. (2011). Endothelial cell activation by antiphospholipid antibodies is modulated by Kruppel-like transcription factors. *Blood*, **117**(23), pp. 6383–6391.

95. Lucitti, J. L., Jones, E. A., Huang, C., Chen, J., Fraser, S. E. and Dickinson, M. E. (2007). Vascular remodeling of the mouse yolk sac requires hemodynamic force. *Development*, **134**(18), pp. 3317–3326.

96. Zarins, C. K., Bomberger, R. A. and Glagov, S. (1981). Local effects of stenoses: increased flow velocity inhibits atherogenesis. *Circulation*, **64**(2 Pt 2), pp. II221–227.

97. Langille, B. L. (1993). Remodeling of developing and mature arteries: endothelium, smooth muscle, and matrix. *J Cardiovasc Pharmacol*, **21**(Suppl 1), pp. S11–S17.

98. Wu, J., Bohanan, C. S., Neumann, J. C. and Lingrel, J. B. (2008). KLF2 transcription factor modulates blood vessel maturation through smooth muscle cell migration. *J Biol Chem*, **283**(7), pp. 3942–3950.

99. Mack, P. J., Zhang, Y., Chung, S., Vickerman, V., Kamm, R. D. and Garcia-Cardena, G. (2009). Biomechanical regulation of endothelium-dependent events critical for adaptive remodeling. *J Biol Chem*, **284**(13), pp. 8412–8420.

100. Sangwung, P., Zhou, G., Nayak, L., et al. (2017). KLF2 and KLF4 control endothelial identity and vascular integrity. *JCI Insight*, **2**(4), pp. e91700.

101. Zhou, G., Hamik, A., Nayak, L., et al. (2012). Endothelial Kruppel-like factor 4 protects against atherothrombosis in mice. *J Clin Invest*, **122**(12), pp. 4727–4731.

102. Yoshida, T., Yamashita, M., Horimai, C. and Hayashi, M. (2014). Deletion of Kruppel-like factor 4 in endothelial and hematopoietic cells enhances neointimal formation following vascular injury. *J Am Heart Assoc*, **3**(1), pp. e000622.

103. Lin, Z., Natesan, V., Shi, H., et al. (2010). Kruppel-like factor 2 regulates endothelial barrier function. *Arterioscler Thromb Vasc Biol*, **30**(10), pp. 1952–1959.

104. Carmeliet, P. (2005). Angiogenesis in life, disease and medicine. *Nature*, **438**(7070), pp. 932–936.

105. Meadows, S. M., Salanga, M. C. and Krieg, P. A. (2009). Kruppel-like factor 2 cooperates with the ETS family protein ERG to activate Flk1 expression during vascular development. *Development*, **136**(7), pp. 1115–1125.

106. Fiedler, U. and Augustin, H. G. (2006). Angiopoietins: a link between angiogenesis and inflammation. *Trends Immunol*, **27**(12), pp. 552–558.

107. Fiedler, U., Reiss, Y., Scharpfenecker, M., et al. (2006). Angiopoietin-2 sensitizes endothelial cells to TNF-alpha and has a crucial role in the induction of inflammation. *Nat Med*, **12**(2), pp. 235–239.

108. Zeng, X. Q., Li, N., Pan, D. Y., et al. (2015). Kruppel-like factor 2 inhibit the angiogenesis of cultured human liver sinusoidal endothelial cells

through the ERK1/2 signaling pathway. *Biochem Biophys Res Commun*, **464**(4), pp. 1241–1247.

109. Kawanami, D., Mahabeleshwar, G. H., Lin, Z., et al. (2009). Kruppel-like factor 2 inhibits hypoxia-inducible-factor-1alpha expression and function in the endothelium. *J Biol Chem*, **284**(31), pp. 20522–20530.

110. De Bock, K., Georgiadou, M., Schoors, S., et al. (2013). Role of PFKFB3-driven glycolysis in vessel sprouting. *Cell*, **154**(3), pp. 651–663.

111. Chen, X. L., Varner, S. E., Rao, A. S., et al. (2003). Laminar flow induction of antioxidant response element-mediated genes in endothelial cells. A novel anti-inflammatory mechanism. *J Biol Chem*, **278**(2), pp. 703–711.

112. Fledderus, J. O., Boon, R. A., Volger, O. L., et al. (2008). KLF2 primes the antioxidant transcription factor Nrf2 for activation in endothelial cells. *Arterioscler Thromb Vasc Biol*, **28**(7), pp. 1339–1346.

113. Kinderlerer, A. R., Ali, F., Johns, M., et al. (2008). KLF2-dependent, shear stress-induced expression of CD59: a novel cytoprotective mechanism against complement-mediated injury in the vasculature. *J Biol Chem*, **283**(21), pp. 14636–14644.

114. Shi, H., Sheng, B., Zhang, F., et al. (2013). Kruppel-like factor 2 protects against ischemic stroke by regulating endothelial blood brain barrier function. *Am J Physiol Heart Circ Physiol*, **304**(6), pp. H796–805.

115. Jain, M. K. and Ridker, P. M. (2005). Anti-inflammatory effects of statins: clinical evidence and basic mechanisms. *Nat Rev Drug Discov*, **4**(12), pp. 977–987.

116. Ridker, P. M., Danielson, E., Fonseca, F. A., et al. (2008). Rosuvastatin to prevent vascular events in men and women with elevated C-reactive protein. *N Engl J Med*, **359**(21), pp. 2195–2207.

117. Sen-Banerjee, S., Mir, S., Lin, Z., et al. (2005). Kruppel-like factor 2 as a novel mediator of statin effects in endothelial cells. *Circulation*, **112**(5), pp. 720–726.

118. Parmar, K. M., Nambudiri, V., Dai, G., Larman, H. B., Gimbrone, M. A., Jr. and Garcia-Cardena, G. (2005). Statins exert endothelial atheroprotective effects via the KLF2 transcription factor. *J Biol Chem*, **280**(29), pp. 26714–26719.

119. Ali, F., Zakkar, M., Karu, K., et al. (2009). Induction of the cytoprotective enzyme heme oxygenase-1 by statins is enhanced in vascular endothelium exposed to laminar shear stress and impaired by disturbed flow. *J Biol Chem*, **284**(28), pp. 18882–18892.

120. Morita, T. (2005). Heme oxygenase and atherosclerosis. *Arterioscler Thromb Vasc Biol*, **25**(9), pp. 1786–1795.

121. Lee, T. S., Chang, C. C., Zhu, Y. and Shyy, J. Y. (2004). Simvastatin induces heme oxygenase-1: a novel mechanism of vessel protection. *Circulation*, **110**(10), pp. 1296–1302.

122. Shields, J. M., Christy, R. J. and Yang, V. W. (1996). Identification and characterization of a gene encoding a gut-enriched Kruppel-like factor expressed during growth arrest. *J Biol Chem*, **271**(33), pp. 20009–20017.

123. Katz, J. P., Perreault, N., Goldstein, B. G., et al. (2005). Loss of Klf4 in mice causes altered proliferation and differentiation and precancerous changes in the adult stomach. *Gastroenterology*, **128**(4), pp. 935–945.

124. Swamynathan, S. K., Katz, J. P., Kaestner, K. H., Ashery-Padan, R., Crawford, M. A. and Piatigorsky, J. (2007). Conditional deletion of the mouse Klf4 gene results in corneal epithelial fragility, stromal edema, and loss of conjunctival goblet cells. *Mol Cell Biol*, **27**(1), pp. 182–194.

125. Swamynathan, S. K., Davis, J. and Piatigorsky, J. (2008). Identification of candidate Klf4 target genes reveals the molecular basis of the diverse regulatory roles of Klf4 in the mouse cornea. *Invest Ophthalmol Vis Sci*, **49**(8), pp. 3360–3370.

126. Segre, J. A., Bauer, C. and Fuchs, E. (1999). Klf4 is a transcription factor required for establishing the barrier function of the skin. *Nat Genet*, **22**(4), pp. 356–360.

127. Hamik, A., Lin, Z., Kumar, A., et al. (2007). Kruppel-like factor 4 regulates endothelial inflammation. *J Biol Chem*, **282**(18), pp. 13769–13779.

128. Adam, P. J., Regan, C. P., Hautmann, M. B. and Owens, G. K. (2000). Positive- and negative-acting Kruppel-like transcription factors bind a transforming growth factor beta control element required for expression of the smooth muscle cell differentiation marker SM22alpha in vivo. *J Biol Chem*, **275**(48), pp. 37798–37806.

129. Liu, Y., Sinha, S., McDonald, O. G., Shang, Y., Hoofnagle, M. H. and Owens, G. K. (2005). Kruppel-like factor 4 abrogates myocardin-induced activation of smooth muscle gene expression. *J Biol Chem*, **280**(10), pp. 9719–9727.

130. Yoshida, T., Gan, Q. and Owens, G. K. (2008). Kruppel-like factor 4, Elk-1, and histone deacetylases cooperatively suppress smooth muscle cell differentiation markers in response to oxidized phospholipids. *Am J Physiol Cell Physiol*, **295**(5), pp. C1175–1182.

131. Yoshida, T., Kaestner, K. H. and Owens, G. K. (2008). Conditional deletion of Kruppel-like factor 4 delays downregulation of smooth muscle cell differentiation markers but accelerates neointimal formation following vascular injury. *Circ Res*, **102**(12), pp. 1548–1557.

132. Feinberg, M. W., Cao, Z., Wara, A. K., Lebedeva, M. A., Senbanerjee, S. and Jain, M. K. (2005). Kruppel-like factor 4 is a mediator of proinflammatory signaling in macrophages. *J Biol Chem*, **280**(46), pp. 38247–38258.

133. Feinberg, M. W., Wara, A. K., Cao, Z., et al. (2007). The Kruppel-like factor KLF4 is a critical regulator of monocyte differentiation. *EMBO J*, **26**(18), pp. 4138–4148.

134. Ghaleb, A. M., McConnell, B. B., Nandan, M. O., Katz, J. P., Kaestner, K. H. and Yang, V. W. (2007). Haploinsufficiency of Kruppel-like factor 4 promotes adenomatous polyposis coli dependent intestinal tumorigenesis. *Cancer Res*, **67**(15), pp. 7147–7154.

135. Zhang, W., Chen, X., Kato, Y., et al. (2006). Novel cross talk of Kruppel-like factor 4 and beta-catenin regulates normal intestinal homeostasis and tumor repression. *Mol Cell Biol*, **26**(6), pp. 2055–2064.

136. Zhang, G., Zhu, H., Wang, Y., et al. (2009). Kruppel-like factor 4 represses the transcription of survivin gene in esophageal cancer cell lines. *Biol Chem*, **390**, pp. 463–469.

137. Takahashi, K. and Yamanaka, S. (2006). Induction of pluripotent stem cells from mouse embryonic and adult fibroblast cultures by defined factors. *Cell*, **126**(4), pp. 663–676.

138. Okita, K., Ichisaka, T. and Yamanaka, S. (2007). Generation of germline-competent induced pluripotent stem cells. *Nature*, **448**(7151), pp. 313–317.

139. Wernig, M., Meissner, A., Foreman, R., et al. (2007). In vitro reprogramming of fibroblasts into a pluripotent ES-cell-like state. *Nature*, **448**(7151), pp. 318–324.

140. Maherali, N., Sridharan, R., Xie, W., et al. (2007). Directly reprogrammed fibroblasts show global epigenetic remodeling and widespread tissue contribution. *Cell Stem Cell*, **1**(1), pp. 55–70.

141. Yet, S. F., McA'Nulty, M. M., Folta, S. C., et al. (1998). Human EZF, a Kruppel-like zinc finger protein, is expressed in vascular endothelial cells and contains transcriptional activation and repression domains. *J Biol Chem*, **273**(2), pp. 1026–1031.

142. McCormick, S. M., Eskin, S. G., McIntire, L. V., et al. (2001). DNA microarray reveals changes in gene expression of shear stressed

human umbilical vein endothelial cells. *Proc Natl Acad Sci U S A*, **98**(16), pp. 8955–8960.

143. Jiang, Y. Z., Jimenez, J. M., Ou, K., McCormick, M. E., Zhang, L. D. and Davies, P. F. (2014). Hemodynamic disturbed flow induces differential DNA methylation of endothelial Kruppel-like factor 4 promoter in vitro and in vivo. *Circ Res*, **115**(1), pp. 32–43.

144. Ohnesorge, N., Viemann, D., Schmidt, N., et al. (2010). Erk5 activation elicits a vasoprotective endothelial phenotype via induction of Kruppel-like factor 4 (KLF4). *J Biol Chem*, **285**(34), pp. 26199–26210.

145. Hsieh, P. N., Zhou, G., Yuan, Y., et al. (2017). A conserved KLF-autophagy pathway modulates nematode lifespan and mammalian age-associated vascular dysfunction. *Nat Commun*, **8**(1), pp. 914.

Chapter 3

Aortic Valve Endothelium Mechanobiology

Rachel L. E. Adams[a,b] **and Craig A. Simmons**[a,b,c]
[a]*Translational Biology and Engineering Program, Ted Rogers Centre for Heart Research, 661 University Ave., 14th Floor, Toronto, ON M5G 1M1, Canada*
[b]*Institute of Biomaterials & Biomedical Engineering, University of Toronto, 164 College St Room 407, Toronto, ON M5S 3G9, Canada*
[c]*Department of Mechanical & Industrial Engineering, University of Toronto, 5 King's College Rd., Toronto, ON M5S 3G8, Canada*
c.simmons@utoronto.ca

3.1 Introduction

3.1.1 The Aortic Valve

The aortic valve functions as a door to the heart, opening to permit the unidirectional flow of blood from the left ventricle to the aorta and closing to prevent blood flow in the reverse direction.

Modern Mechanobiology: Convergence of Biomechanics, Development, and Genomics
Edited by Juhyun Lee, Sharon Gerecht, Hanjoong Jo, and Tzung Hsiai

ISBN 978-981-4800-58-7 (Hardcover), 978-0-429-29483-9 (eBook)
www.jennystanford.com

Anatomically, the aortic valve comprises three leaflets: the left coronary (LC), the right coronary (RC), and the noncoronary (NC). During the cardiac cycle, the leaflets open for ventricular systole (ejection) and close for ventricular diastole (filling). The leaflets of the human aortic valve have sophisticated tissue architecture; each leaflet can be segmented into three axial layers of tissue that have unique extracellular matrix (ECM) properties but are structurally and biologically integrated. The fibrosa layer, on the aortic side of the valve, bears the majority of the mechanical load from pressure in the aorta [1]. A high concentration of stiff, circumferentially-aligned type I collagen in this layer strengthens the ECM against the deformation of the valve during diastole [2]. The ventricularis layer, on the ventricle side of the valve, has radially aligned elastin, which creates a pliable ECM able to respond to the continuous opening and closing of the valve [3, 4]. The spongiosa layer is situated in the middle of the leaflet, between the fibrosa and the ventricularis. Composed mainly of glycosaminoglycans and proteoglycans [5], the spongiosa acts as a lubricant between the layers of opposing stiffness—stable enough to withstand the hemodynamic forces of the active circulation and compliant enough to continually open and close [6].

A subset of the population (1%–2% [7, 8]) is born with a bicuspid aortic valve (BAV), in which the aortic valve develops with two of the three leaflets fused, resulting in only two leaflets in total. BAVs are the most common congenital heart defect [9], and the most common type of fusion is classified as Type I, with fusion of the LC and RC leaflets [10, 11]. BAVs often develop a raphe, a ridge formed on the fibrosa side of the leaflet, generally found on the fused leaflet of BAVs of unequal-sized cusps; raphes can contribute to the increased mechanical stiffness of the tissue [11]. BAVs can also have varying degrees of eccentricity, which describes the degree to which the valve orifice is circular versus elliptical. Due to their altered anatomy, BAVs often experience abnormal hemodynamics, which is thought, but not yet proven, to predispose them to disease development [12].

3.1.2 Aortic Valve Cell Types

The ECM of the aortic valve leaflets is sparsely populated by a heterogeneous population of cells called valve interstitial cells (VICs). In healthy valves, the VIC population mainly consists of quiescent fibroblasts [13] that synthesize the ECM components (collagen, elastin, laminin, fibrin, and glycoaminoglycans) [14], progenitor cells [15], and a small population (<5%) of myofibroblast cells that express α-smooth muscle actin (α-SMA). Similar to other blood-contacting surfaces in the circulation, a monolayer of valve endothelial cells (VECs) lines both the aortic surface and the ventricular surface of the aortic valve leaflets. The collective role of these cells is similar to that of vascular endothelial cells: regulating valve permeability [16], expressing selectin molecules for cytokine and inflammatory cell adhesion [17, 18], and sensing hemodynamic cues from the circulating blood [19]. It is also likely that bidirectional paracrine signaling between VECs and VICs contributes to aortic valve homeostasis, the disruption of which can predispose local environments to pathogenesis [20, 21].

3.1.3 Calcific Aortic Valve Disease

Calcific aortic valve disease (CAVD) is the most prevalent disease related to heart valves, affecting over 25% of people over the age of 65 [22]. The disease encompasses an early stage clinical presentation called aortic sclerosis, which is an active inflammatory process of lipid deposition and regional lesion formation in the subendothelial space of the fibrosa [17, 18]. The later stage of the disease is referred to clinically as aortic stenosis, which can involve tissue thickening (fibrosis) and deposition of calcium nodules (calcification) [17, 23], which lead to valve orifice narrowing. Functionally, this causes impairment of leaflet motion and obstruction of blood flow to the aorta. If left untreated, a stenotic aortic valve can lead to increased transvalvular pressure gradients, left ventricular pressure overloading, left ventricular hypertrophy, and myocardial fibrosis, ultimately leading to heart failure. CAVD is a major risk factor of heart disease, myocardial infarction [24, 25], and stroke and is thus a contributor to the current and projected major causes of death worldwide [26, 27]. With no current therapeutic treatment, progression of CAVD ultimately requires expensive and invasive heart valve transplants that carry the risk of stroke [28].

It is not known what causes CAVD. Initially considered a degenerative, passive, or wear-and-tear disease, CAVD is now recognized to be a molecularly regulated process [29]. CAVD is believed to involve VEC dysfunction [30] and phenotypic changes in VICs that lead to maladaptive matrix remodeling [31]. Fibrosis, the thickening and scarring of tissue, is caused by VIC differentiation to myofibroblasts, expressing characteristic α-SMA [32, 33], an actin isoform that has been shown to enhance contractile activity [34]. Calcification is the ectopic deposition of calcium nodules in the valve leaflets. This pathology is classified as either dystrophic (calcium crystal nucleation in zones of necrotic tissue) or osteogenic (deposition of mineralized matrix and formation of bone) [35]. The latter is an active process that involves osteoblast-like cells with markers of bone, such as osteopontin and osteocalcin, as well as the presence of bone morphogenic proteins (BMPs). Within valve pathogenesis, it is not known whether myofibrogenesis precedes osteogenesis or whether these diseases are mutually exclusive events.

BAVs develop CAVD earlier and to a greater extent than tricuspid aortic valves (TAVs) [1, 4]. Consequently, while approximately 2% of the population is born with a BAV, this population constitutes 50% of all aortic stenosis cases [36–38]. While some of the underlying genetics that cause BAVs are understood and likely contribute to accelerated CAVD, the altered anatomical structure of BAVs and resultant hemodynamics are also hypothesized to play a compelling role [39].

3.1.4 Aortic Valve Mechanics

Cells within the aortic valve are exposed to physical stresses, including the inherent stiffness of the leaflet tissue, stretch when the leaflets are closed and withstanding the back pressure of blood in the aorta during diastole, and shear stress exerted on the endothelium by blood flowing around the open leaflets during systole. Valve cells sense and respond to these biomechanical stimuli, eliciting biological responses that can contribute to valve homeostasis and disease in a process known as mechanobiology.

Among the biomechanical regulators of aortic valve endothelial biology, shear stress (the tangential component of the frictional force applied by flowing viscous fluid on a surface) is likely the most potent. In the aortic valve, shear stress is applied by the flow of viscous blood on the leaflet, acting along the tissue endothelium

(Fig. 3.1A). Both sides of the aortic valve endothelium are exposed to shear stress, but the dynamics of blood flow and its interaction with the valve tissue throughout the cardiac cycle result in each side of the valve endothelium experiencing unique shear stress magnitudes, directions, and time courses. Blood flow can also change over time due to variations in blood pressure, heart function, and vascular dilation. In this way, VECs act as the gatekeeper for evolving shear stress mechanotransduction. However, despite growing evidence that shear stress regulates valve homeostasis [40], shear stress–regulated pathways are not fully defined. It is expected that mapping out shear stress–induced VEC signaling pathways is integral to understating aortic valve homeostasis as well as its susceptibility to disease development.

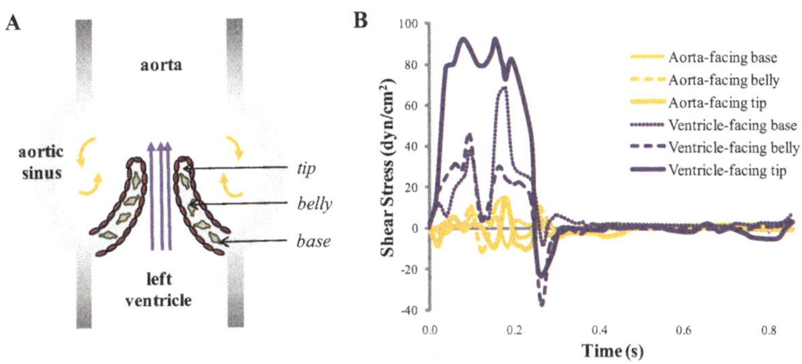

Figure 3.1 Side- and region-dependent blood flow-induced shear stresses on the leaflets of the aortic valve. (A) The aortic valve, situated between the left ventricle and aorta, is exposed to flow-induced shear stress on both sides of the valve leaflets. (B) A computational estimation of aortic valve shear stress waveforms demonstrating high magnitude, largely unidirectional shear stress on the ventricle-side, and lower magnitude, oscillatory shear stress on the aorta-side. Waveforms adapted from Fig. 6 of Ref. [71].

3.1.5 The Role of Shear Stress in the Aortic Valve Endothelium

The fluid mechanics of flowing blood, referred to as hemodynamics, fluctuates in magnitude and velocity depending on cardiac output and local vessel geometry. At the cellular level, the endothelium is subjected to laminar blood flow, of which the tangential component of the frictional force is called wall shear stress (WSS). Shear stress in the aortic valve cannot be directly measured; rather, it is calculated from known local parameters, such as velocity, viscosity, and geometry. The aortic valve experiences different hemodynamics throughout the cardiac cycle, and the leaflets are exposed to different stresses on their opposite sides; as a result, shear stress values are temporally and spatially dynamic and are consequently challenging to calculate.

The relationship between hemodynamic shear stress and endothelial cells is important. In the vasculature, the endothelium is responsible for transmitting hemodynamic cues from flowing blood to underlying tissue [41]. It has been shown in vitro that vascular endothelial cell gene expression, proliferation, and monocyte adhesion are regulated by atheroprone flow characteristics, such as low shear stress and oscillatory fluid behaviors [42]. While it might seem that vascular and valvular endothelial cells would have similar in vivo responses to mechanical stimuli, the aortic valve differs in architecture and movement, which results in the generation of unique shear stress waveforms, giving rise to a medical need for the specific evaluation of aortic valve mechanobiology.

To this end, innovative research has tried to delineate the role of shear stress in CAVD, with a focus on:

- Defining the shear stress waveforms for both TAVs and BAVs
- Studying the response of endothelial cells to shear stress
- Mapping out shear stress–regulated pathways as they relate to aortic valve homeostasis and disease

This chapter focuses on the mechanobiology of the aortic valve endothelium, and progress toward each of these goals is outlined in the following sections. It summarizes advancements in the understanding of the role of shear stress in valve endothelial biology and identifies areas of research that require further study.

3.2 Shear Stress Waveforms of Aortic Valves

Cardiac hemodynamics have been studied over many decades, with a focus on flow through the aortic valve and on to the ascending aorta. These preliminary studies served to create diagnostic tools for non-invasive detection of aortic stenosis or the presence of a congenital BAV [43, 44]. Over time, techniques have evolved to examine hemodynamics and its role in aorta pathology [45], such as predisposition to aneurysmal development [46] and atherosclerosis [47]. Cardiac hemodynamics related to the aortic valve initially focused on macroscopic parameters, such as jet velocities, leaflet kinematics, and BAV eccentricities. Only recently have there been studies looking at the shear stress acting on aortic *valve* endothelium specifically.

The shear stress acting on the aortic valve endothelium deserves attention for two reasons. First, aortic valve disease is a different pathology than disease of the aorta [1, 48]. Second, endothelial cells of the valve and aorta respond differently to mechanical stimuli. For example, porcine VECs show different transcriptional profiles than aortic endothelial cells and proliferate more rapidly on collagen or gelatin, suggesting different mechanosensing properties [49]. Further, cell adhesion, which regulates responses to mechanobiological stimuli such as shear stress, can also differ between vascular and valvular endothelium [50]. It has been hypothesized that differences between vascular endothelial cells and VECs stem, at least in part, from differences in the local shear stresses they experience in vivo [51]; in other words, vascular and valvular endothelial cells may be programmed differently by their microenvironments.

3.2.1 Aortic Valve Shear Stress Waveforms Are Estimated

Velocity profiles through the aortic valve are used to estimate the WSS experienced by the valve endothelium. Both TAV and BAV velocity profiles have been characterized in vivo using techniques such as echocardiography [43, 44], magnetic resonance imaging (MRI) [46, 47], and flow visualization [52]. These qualitative studies correlated velocity profiles with sites of disease in the aorta in an effort to characterize disease-prone versus disease-protective hemodynamic flow. However, absolute quantification of shear stress on the aortic valve is needed in order to model physiological flow in ex vivo and in vitro systems.

Researchers use both experimental and theoretical approaches to quantify velocity profiles. Flow visualization is an experimental technique that utilizes the basics of medical imaging to create images of, and track fluid flow in, the aortic valve environment. Examples of this are laser Doppler velocimetry (LDV) [53, 54], which measures reflecting light waves, and particle image velocimetry (PIV) [55, 56], which traces the movement of microspheres. Although these techniques use physiologically relevant models, they lack resolution at the wall of the tissue where the WSS would be applied. Alternatively, 4D flow MRI is a highly valuable technique that allows evaluation of multidirectional blood flow, including details of helical and vortical-type velocities [57]. However, it is limited in the characterization and resolution of spatial and temporal changes and so fails to capture the complexity of the WSS throughout the cardiac cycle [58].

Theoretical estimates of WSS utilize finite element analysis to create models of the aortic valve structure and computational fluid dynamics to capture the dynamics of flowing fluid. Used together, these techniques can model the fluid dynamics within a defined structure, such as the aortic valve, though they lack the impact of the fluid on the structure itself. To reconcile this, fluid-structure interaction (FSI) is a tool developed to calculate the forces acting on a solid, deformable structure, such as the aortic valve, *in contact* with a flowing fluid, such as blood. This technique accounts for the transfer of momentum between the leaflets and blood flow (i.e., their interaction) [59, 60]. The challenges with this method are the difficulty of constructing an accurate model of the tissue leaflets within the aortic sinus as well as modeling the complexity of physiological blood flow. Fortunately, emerging vascular studies have combined 4D MRI with computational methods to achieve better estimations of WSS of the aorta [61, 62]. Therefore, the tools are available to apply the same methodology for aortic valve WSS determination.

3.2.2 Aortic Valves Have Side-Specific Shear Stress Waveforms

Endothelium at different locations on the aortic valve experiences different magnitudes and frequencies of shear stress during the cardiac cycle. FSI modeling of the combined left ventricle and aortic valve demonstrates that, during systole, contraction of the left ventricle generates flow [63] that creates a high pressure gradient across the valve and forces the valve to open. High-velocity fluid flows in the axial direction across the aortic valve aperture into the aortic sinus and proximal aorta [64]. Leaflet movement is governed by the motion of the fluid [59] and leaflets respond to the systolic jet with only a small degree of asymmetry [63]. This high-velocity jet has some orientation toward the aortic walls, which creates vortices in the aortic sinus [65]. As the cardiac cycle transitions from systole to diastole, a drop in transvalvular pressure causes reversal of blood flow, aortic valve leaflet closure, and a corresponding decrease in shear stress. This cardiac cycle results in unique microenvironments and shear stress profiles on the aorta-facing (aorta side) and the ventricle-facing (ventricle side) endothelium. Analysis of the temporal and spatial differences of these waveforms can provide a more accurate picture of WSS throughout the cardiac cycle and throughout different areas of the leaflets.

Hemodynamic flow on the ventricle side of the valve is considerably different than flow on the aorta side. During systole when the valve is fully opened, ventricle-side velocity vectors are parallel and uniform, whereas aorta-side vectors are characterized as vortical structures with counter-rotating flow profiles [66]. The first study to use FSI for quantification of leaflet WSS explored the spatial variation of WSS in the radial direction [60]. With a focus on systolic blood flow, the authors demonstrated that blood exerts a high, unidirectional WSS (20–80 dyn/cm^2) on the ventricle side of the opened valve leaflet, with the WSS magnitude decreasing from base to tip. They also demonstrated that the aorta-side endothelium experiences a low-magnitude (<5 dyn/cm^2) WSS with a high degree of oscillation in the radial direction.

Not all studies have arrived at the same conclusion about the magnitude of shear stress on the aorta-side endothelium. In a study looking at excised aortic valves, measurements of the WSS on the leaflet surfaces showed that the aorta side of the valve experiences a high, unidirectional WSS of only 20 dyn/cm^2 at peak systole [67]. To measure this, normal porcine aortic valve leaflets were placed in mounting chambers that mimicked the aortic sinus geometry. Valves were then subjected to pulsatile flow, and time-varying WSS were estimated using liquid crystal display. The result is contrary to commonly modeled [60, 66, 68] low-oscillatory aorta-side WSS and could be explained by the design of the mounting chambers. Aortic valve leaflets do not open 90° during systole, resulting in entrapment of blood in the aortic sinus. During systole, aorta-side sinus vortices are induced by the interaction between forward flow and fluid in the sinus [69, 70] and are affected by the degree to which the valve opens during systole and how much of the systolic jet directly hits the sinus wall. The strong adjacent flow of the transvalvular jet then induces the vortex in the sinus. The strength of this interaction results in a unidirectional vortex that is not believed to reverse, or oscillate, in the radial direction. Therefore, perhaps the difference in magnitude and flow profiles for the aorta side is due to subtle variations in the alignment of the excised aortic valves in the sinus chamber apparatus. Either way, it is clear that small changes in geometry or hemodynamics can alter WSS profiles, highlighting the challenge for causal diagnosis and the need for physiologically representative WSS waveforms.

3D models may help to create more physiologically accurate flow profiles. A more recent FSI model evaluated blood flow in three dimensions with a systolic velocity jet into the aortic sinus cavity [71]. This model included spatial characterization of WSS in the radial and circumferential directions. As expected, radial WSS was found to be more dominant than circumferential WSS. Ventricle-side radial WSS waveforms were defined for the tip, belly, (46–70 dyn/cm^2), and base regions (93 dyn/cm^2) (Fig. 3.1B). Furthermore, low-amplitude and bidirectional WSS were estimated for the aorta side of the leaflet. The same researchers more recently updated their model to include coronary blood flow [72]. They demonstrated that coronary flow suppressed vortex development in the coronary sinuses, generating an increase in WSS magnitude and spatially dependent flow directionality on the aorta side of the leaflet. As coronary blood flow only exists for left- and right-coronary leaflets, this is the first study to highlight potential differences in WSS magnitude and directionality between the three leaflets, which is indeed relevant since CAVD often occurs first on the NC leaflet [1].

3.2.3 Bicuspid Aortic Valves Have Abnormal Shear Stress Waveforms

The hemodynamics of the BAVs have been studied extensively for their association with, and effect on, aortopathies [72–79]. Modeling of TAV and BAV structures reveals that when the valve opens during systole, BAVs open asymmetrically, which results in an elliptical opening for blood flow out of the heart [80]. Echocardiographic evaluation of hemodynamic flow profiles of BAVs demonstrates the presence of higher aortic peak velocities at systole [73, 74] that have nested, helical flow [46, 81]. 4D MRI modeling demonstrates that these abnormal flow patterns generate higher WSS at the aortic wall. The locations of higher WSS also correlate with the diameter of the ascending aorta [45], suggesting that the endothelium of the aorta is also mechanically responsive.

At the aortic valve, the impact of altered BAV flow patterns on valve pathogenesis has only recently been studied. This is likely because mechanically regulated CAVD is commonly thought to be secondary to the genetic basis for disease development. Also, the WSS of BAVs is more elusive to obtain, as a result of the different types of BAV fusion (LC/RC, RC/NC, or LC/NC) and the range of eccentricity that BAVs can exhibit [10, 46, 75]. BAVs have inherent morphological stenosis as a result of their fusion, which creates disturbed hemodynamics in vivo. In addition, BAVs undergo excessive folding during the cardiac cycle [52]. Although there is indisputable evidence that BAVs are caused by genetic mutations [9, 82–84], it is not known what contribution altered hemodynamics play on the accelerated aortic valve disease development this population typically experiences.

In vitro characterization of BAV hemodynamics has shown that flow across the leaflets of BAVs is more disturbed than that of TAVs [65]. Using porcine leaflets in an ex vivo pulsatile flow loop with idealized sinus geometry, the authors made artificial BAVs from two TAV leaflets sutured together. Comparing TAVs to BAVs, they used PIV to characterize flow downstream of the leaflets and LDV tracking to estimate the WSS on the central belly of the aorta side of the leaflets. The model demonstrated that BAVs have skewed forward flow, offset to the central sinus axis. Compared to the fused leaflets, the nonfused side of the BAV had higher vortices in the sinus and higher aorta-side peak WSS (36 dyn/cm^2 compared to 20 dyn/cm^2). Shear stress was also shown to be in opposing directions on the fused versus nonfused leaflets. Interestingly, TAV flow and peak WSS were shown to be similar to those for the BAV fused leaflets (19.6 dyn/cm^2), though the authors acknowledged this could be due to the mild stenosis of the TAV model. In a subsequent study, this group used the same experimental setup to further characterize the flow in two BAV models with different degrees of stenosis [56].

Consistent with their previous study, they showed that higher jet velocities govern vortical development in the nonfused leaflet sinus, resulting in a larger magnitude of shear stress on the tissue. They further showed that this occurs to a greater extent in Type-1 BAVs (with raphe) that have a higher degree of eccentricity and an asymmetrical aortic root opening, leading to flow profiles that are closer to clinical measurements of BAV flow [46].

Numerical models have also been used to characterize fluid flow in BAVs. In a very elegant design, Marom et al. used a fully computational 3D FSI model to obtain physiologically relevant asymmetric BAVs [85]. The incorporation of improved valve mechanics (calculating strain as a function of pressure and correctly oriented tissue components) and geometry (designing fused leaflets with or without raphes) in the model improved flow predictions. In alignment with previous studies, their model found that peak systolic flow through asymmetric BAVs causes larger vortices and higher shear stress on the aorta side of the BAV cusps compared with TAVs. However, the peak shear stresses were found to be an order of magnitude higher than what has been predicted in in vitro modeling. Specifically, the peak systolic WSS on the aorta-side TAV was 280 dyn/cm^2 and for symmetrical and asymmetrical BAV models, 980 and 4800 dyn/cm^2, respectively. Similarly, the peak WSS at tip of the ventricular side was calculated to be 420, 1430, and 4800 dyn/cm^2 in TAV, BAV (symmetrical), and BAV (asymmetrical), respectively. This model highlighted that the higher the degree of stenosis, the greater the WSS at the tip of the leaflets. Certainly, aortic valve leaflets experience a gradient of shear stress, both spatially and temporally. Whether exact shear stress profiles are necessary when studying endothelial responses in vitro or whether there are different effects on the endothelium biology between WSS magnitude, directionality, and oscillation frequency remains incompletely understood. Borrowing from vascular mechanobiology [86], it might be possible that simplified elements of WSS can be extracted from the waveforms for use in measuring endothelial responses to their local WSS environments.

As described earlier, Chandra et al. created a 2D FSI model that predicted full WSS waveforms throughout the cardiac cycle [60]. In addition to TAV waveforms, they studied Type-1 BAV models with different eccentricities. Contrary to PIV studies [56, 65], they found that higher WSS occurs on the fused leaflets and this was independent of the BAV model. It has yet to be shown whether calcification in BAVs occurs to a greater extent on the nonfused leaflets or the fused leaflets, but it is likely that the difference in disease progression could be due, in part, to the differences in local WSS at the endothelial wall. Before a solution can be found, the problem must be fully defined. In this case, discerning which BAV models best describe the hemodynamics will be necessary to establish physiologically relevant disease models. It is accepted that low, oscillatory WSS is disease permissive, but it is also possible that abnormally higher WSS can upregulate pathological signaling cascades, possibly seen more in patients with BAVs.

3.3 Valve Endothelial Response to Shear Stress

In the vasculature, endothelial cells are regulated by shear stress [87–89]; the transmission of such stimuli and their role in vascular homeostasis and disease are widely studied [19, 90, 91]. Undisturbed laminar shear stress is known to protect the endothelial layer by regulating vasodilation [92–94], downregulating adhesive molecules [95], and enabling antioxidant gene expression [96], whereas disturbed flow has the opposite effect, indicating that vascular endothelial cells distinguish different types of shear stress stimuli [97, 98]. Progress toward studying the phenotypic plasticity of VECs in response to shear stress is well underway [51, 99] and is necessary to understand protective and dysfunctional valvular endothelial responses in aortic valve biology.

3.3.1 Devices Designed for Studying VEC Response to Shear Stress

Advancements in design have led to novel devices for cultures of VECs that closely mimic in vivo conditions. Microfluidic devices allow for complex cell arrangement, minimal reagent use, and high-throughput data generation [100, 101]. These technologies can be used to mimic the microvasculature and can be adapted for the aortic valve [21]. As a macrofluidic device, cone-and-plate devices (Fig. 3.2A) can be used to deliver pulsatile and oscillatory waveforms on cells [97, 102]. This system can also be adapted for ex vivo applications to apply waveforms on aortic valve tissue endothelium, via one-sided [103] or double-sided cone-and-plate bioreactors [104]. Such devices permit the study of mechanically regulated VEC morphology and gene regulation.

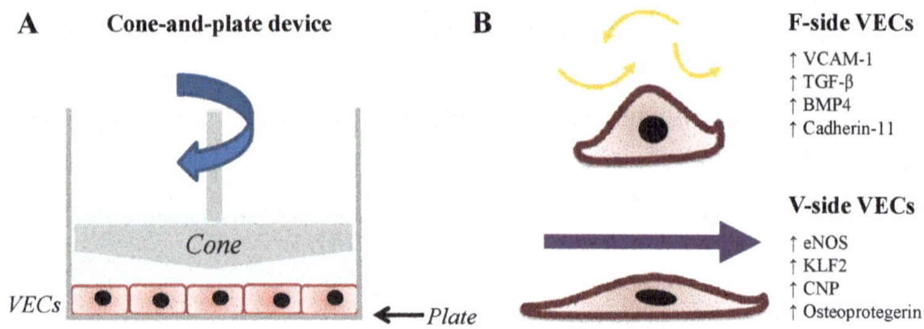

Figure 3.2 Valve endothelial cells (VECs) have unique phenotypes based on the type of shear stress they are exposed to. (A) Cone-and-plate devices can be used to expose VECs to physiological shear stress waveforms in vitro. (B) The plasticity of valve endothelial cell phenotype is affected by the type of shear stress: VECs exposed to low, oscillatory shear stress express pathological markers VCAM-1, TGF-β, BMP4, and cadherin-11, whereas VECs exposed to high, unidirectional shear stress are elongated, with upregulation of the protective factors eNOS, KLF2, CNP, and ostoprotegerin.

Macro- and microfluidic devices [21] have also been adapted to study shear stress–induced endothelial mechanobiology and its effect on VIC pathogenesis. These devices are valuable because they incorporate delivery of physiological fluid flow and provide a 3D matrix for the correct VEC/VIC orientation to allow for paracrine signaling between cell types. An early aortic valve coculture model used type 1 collagen gel constructs for culturing porcine VICs [20]. Embedded VICs exposed to 20 dyn/cm^2 of shear stress had increased proliferation and α-SMA expression, which was modulated when cells were cultured with a VEC monolayer that was instead exposed to the shear stress stimulus. In another aortic valve 3D microfluidic model, methacrylated gelatin hydrogels were used to obtain tunable elastic moduli via in situ polymerization of the 3D hydrogel [21]. VICs encapsulated in the hydrogel were separated from VEC monolayers by a porous membrane. VECs suppressed pathological activation of VICs to myofibroblasts contained in hydrogels of physiologically relevant stiffness. This effect was enhanced through exposure of the

VEC monolayer to 20 dyn/cm^2 shear stress. Such novel devices are essential to further the mechanistic insight into mechanoregulated aortic valve disease and, specifically, to highlight the specific role the valve endothelium plays in modulating these responses. However, the question remains as to whether full aortic valve waveforms are necessary for testing in in vitro applications or whether simplified waveforms are enough to elicit the same phenotypic changes of cells as has been shown in a vascular application [86].

3.3.2 VEC Phenotype Is Shear Stress Regulated

Shear stress regulates VEC phenotype: flowing blood is sensed by VEC membrane mechanosensors, stimulation of which can cause a cascade of signaling events that affect gene transcription or cause a conformational change that alters the cytoskeletal structure of the cell. Typically, vascular endothelial cells reorganize their cytoskeleton and elongate into monolayers that align parallel with the direction of flow-induced shear stress [41]. There is conflicting evidence as to whether VECs do the same. Butcher et al. found that VECs subjected to 20 dyn/cm^2 shear stress aligned perpendicular to flow [105]. In contrast, others have shown that VECs under the same shear stress application align parallel to flow [106]. In vivo evaluation has suggested that there are regional differences with VEC alignment throughout the valve that correlate with ECM topography, that is, due to collagen fiber alignment [107], or perhaps VECs are acutely responsive to circumferential versus radial shear stress waveforms at the surface of the endothelium [71]. The particulars of whether VECs are preprogrammed to align differently than vascular endothelial cells, or whether aorta-side and ventricle-side VECs align differently in response to flow, are unknown. Presumably, accurate cell alignment is necessary for studying VEC-related pathways in vitro.

VECs on the surface of the leaflets act as a barrier between the surrounding blood and tissue and, as such, play a role in immune cell adhesion and inflammation. The surface cell adhesion molecules intracellular adhesion molecule-1 (ICAM-1) and vascular cell adhesion molecule-1 (VCAM-1) are transmembrane proteins typically expressed by endothelial cells that act to bind interleukins and other adhesion cells of the immune response as part of transmigration across the endothelium. VECs routinely express basal levels of ICAM-1, but in diseased aortic valves, both ICAM-1 and VCAM-1 are upregulated [18]. Interestingly, this inflammatory response has shown to be mechanically induced [108]. Using excised porcine aortic valves, the aorta side and the ventricle side were exposed to their normal physiological shear stress waveforms (oscillatory and unidirectional high magnitude, respectively) for 48 hours. The authors reported no detectable ICAM-1 or VCAM-1 in immunohistologically stained tissue. However, when each leaflet side was exposed to altered shear stress waveforms (unidirectional high-magnitude shear stress on aorta side and oscillatory shear stress on the ventricle side), the aorta-side valves had positive ICAM-1 and VCAM-1 staining. The stimulation of an inflammatory phenotype only on the aorta side, and distinctly in response to an altered shear stress insult, suggests that aorta-side VECs are pathosusceptible and disease initiates under abnormal hemodynamics. Further, it can be hypothesized that physiological low-magnitude oscillatory shear stress preconditions aorta-side VECs to develop dysregulated phenotypes.

3.3.3 Side-Dependent Hemodynamics Correlate with Side-Specific Phenotypes

The aortic valve experiences side-specific shear stress waveforms that correlate with side-specific VEC phenotypes (Fig. 3.2B). Simmons et al. were the first to characterize the side-specific gene expression profiles of aorta- and ventricle-side VECs [109]. Using RNA amplification and cDNA microarray analysis, they compared the expression profiles of aorta-side and ventricle-side VECs isolated from normal porcine aortic valves. Of the 584 genes differentially regulated, they found that aorta-side VECs had lower expression of multiple inhibitors of calcification, such as osteoprotegerin, C-type natriuretic peptide (CNP), and parathyroid hormone and higher expression of transcripts related to bone formation, such as bone morphogenic protein 4 (BMP4) and cartilage. They also found that the aorta-side VECs expression profile was non-inflammatory and antioxidative. Collectively, the distinct VEC phenotypes suggest the aorta-side environment is permissive to calcification. Furthermore, this work supports the hypothesis that the side-specific VEC phenotype heterogeneity and pathosusceptibility are at least in part regulated by exposure to differential shear stress waveforms in vivo.

Subsequent studies that performed transcriptional profiling of VECs exposed to laminar shear stress demonstrated how shear stress regulates gene expression to promote protective endothelial phenotypes. In a microarray analysis of porcine VECs exposed to 20 dyn/cm^2, 545 genes were found to be upregulated and 311 genes downregulated compared to cells in static conditions [110]. Of these mechanosensitive proteins, unidirectional shear stress was found to inhibit markers of chondro-/osteogenic differentiation, such as BMP4 and the cell junction protein cadherin-11. In a subsequent study, human umbilical vein endothelial cells (HUVECs) were exposed to aorta-side and ventricle-side shear stress waveforms and the expression of key regulators of pathogenesis was measured. The expression of the transcription factors Krüppel-like factor 2 (KLF2), a known inhibitor of BMP4, was shown to be more highly expressed in HUVECs exposed to ventricle-side waveforms. In addition, the protein nephroblastoma overexpressed, an inhibitor of osteogenesis, was also shown to be upregulated in HUVECs exposed to ventricle-side waveforms. Contrastingly, MCP1,[1] a protein implicated in atherosclerosis [111], was found to have a significantly lower expression [112]. Taken together, they support the hypothesis that the ventricle-side shear stress waveforms are a protective modulator of ventricle-side endothelial biology.

Other studies have revealed shear stress–sensitive messenger RNAs (mRNAs) and microRNAs (miRNAs) in VECs from noncalcified human specimens [106]. Using in vitro platforms, aorta-side and ventricle-side VECs were exposed to oscillatory (±5 dyn/cm^2) or unidirectional (20 dyn/cm^2) flow, representing aorta-side and ventricle-side shear stress waveforms, and gene levels were characterized via microarray analysis. Additionally, both aorta-side and ventricle-side VECs were exposed to oscillatory or unidirectional shear stress, to test for characteristic side-dependent mRNA differences. Cell-dependent mRNA differences were *not* detected in aorta-side versus ventricle-side VECs, further supporting the hypothesis that differences in expression seen previously [109] are the result of differences in spatial hemodynamics as opposed to inherent

[1]Monocyte chemoattractant protein 1.

differences in aorta- and ventricle-side endothelial cell types. When comparing cultures of aorta-side VECs exposed to oscillatory flow and ventricle-side VECs exposed to unidirectional flow, representing in vivo conditions, >300 genes were upregulated and >700 genes were downregulated under oscillatory flow. The expression of known shear stress–responsive genes *KLF2*, endothelial nitric oxide synthase (*eNOS*), and *BMP4* was validated, in combination with other genes related to apoptosis, proliferation, and development. For example, the gene *Nfkbia* was downregulated under oscillatory flow, which is physiologically relevant since it is known to sequester NF-κB[2] to inhibit apoptosis. This data analysis also identified 30 shear stress–sensitive miRNAs, 3 of which were confirmed with real-time polymerase chain reaction: miRNA-486-5p, miRNA-139-3p, and miRNA-187. Such novel work lays the groundwork for the discovery of complete mechanosensitive pathways and potential target therapeutics. For example, miRNA-181b, which was later shown to be upregulated in human VECs exposed to low, oscillatory shear stress, has been found to directly target TIMP3 to regulate ECM degradation by VECs [113]. The functional roles of these and other miRNAs in aortic valve homeostasis or disease pathogenesis is complex but is deservedly a quickly evolving field of research in aortic valve biology.

There is a counterargument to the shear stress regulation of side-specific endothelial biology [114]. In a study that used a scanning ion conductance microscope to study intact porcine aortic valve leaflets, the authors found that flow-induced cell deformation caused aorta-side VECs to be more compliant, expressing lower levels of f-actin cytoskeleton filament. This side-specific mechanical stiffness persisted in vitro, under static and physiological flow conditions, even when aorta-side VECs were treated with ventricle-side waveforms and ventricle-side VECs with aorta-side waveforms. VEC stiffness was not governed by the stiffness of the underlying tissue, nor was it regulated by flow patterns applied to the endothelium; rather VECs modulated their own compliance. This study suggests that the differential biomechanics of the valve endothelium is not shear stress regulated, more that shear stress regulates the biochemical responses, and this is in a side-specific manner.

[2]Nuclear factor kappa light chain enhancer of activated B cells.

3.4 Shear Stress-Regulated Mechanisms of Valve Homeostasis and Disease

Fibrotic and calcific lesions in CAVD occur preferentially in the fibrosa layer of the leaflet tissue. It is hypothesized that flow-induced shear stress localized to the valve endothelium modulates VEC paracrine signaling to underlying VICs in a process referred to as mechanotransduction, and this occurs in a side-dependent manner. Mapping out shear stress–regulated pathways is essential for the development of both pharmacological interventions of disease and for the creation of diagnostic tools for disease prevention. Therefore, beyond studying the response of endothelial cells to shear stress, there is a growing field of research that has focused on the integrated VEC response and changes to neighboring cells.

3.4.1 Endothelial to Mesenchymal Transformation

During valvulogenesis in the embryo, valvular endothelial cells can undergo a process known as endothelial-to-mesenchymal transformation (EndMT), whereby VECs migrate and invade the tissue, losing their VEC phenotype but gaining a myofibroblast-like one [115]. This process populates the immature valve cushions with active, synthetic cells, eventually leading to mature and functional valve tissue. This same phenomenon can re-emerge in adult aortic valves, possibly as a cell-renewal mechanism, but also pathologically: an increase in synthetic, myofibroblast-like cells can lead to fibrosis. EndMT has previously been known to be induced by cytokines and growth factors [116]; however, recently, EndMT has been shown to be shear stress regulated as well [117]. Using a 3D parallel plate bioreactor, porcine VECs were exposed to unidirectional shear stress (2, 10, or 20 dyn/cm^2) or oscillatory shear stress (±2, ±10, or ±20 dyn/cm^2), compared with static controls. They demonstrated approximately three times higher invasion of cells into the 3D matrix in conditions of lower-magnitude oscillatory shear stress (±2 and ±10 dyn/cm^2), and this was complemented by the upregulation (gene and protein expression) of the EndMT markers α-SMA and Snail and a loss of VEC adhesion and cell-cell junction proteins VCAM-1 and VE-cadherin. While EndMT has also been shown to be mechanically regulated via cyclic strain [118] and stiffness [119], it is likely that some combination of these physiological mechanical stimuli is responsible for the occurrence in vivo that can play a role in initiating CAVD.

3.4.2 eNOS, Nitric Oxide, Notch1, and Cadherin-11

It is hypothesized that VECs protect against aortic valve disease development via paracrine signaling of nitric oxide (NO) to VICs (Fig. 3.3). NO is a small signaling molecule synthesized by the enzyme eNOS in endothelial cells [120]. This signaling pathway regulates homeostatic events in the vasculature, such as vasodilation and vascular remodeling [121–123]. The NO pathway has been shown to be protective in the aortic valve; valve endothelial-derived NO reduces matrix calcification in porcine VICs in pro-osteogenic conditions [124], and mice lacking eNOS develop aortic sclerosis [125]. As in the vasculature [93, 126], the NO pathway has been shown to be mechanically regulated in the aortic valve. VEC-derived NO expression was found to be shear stress- and side-dependent in porcine aortic valves exposed to side-specific physiological waveforms [127]. Using pharmacological donors and inhibitors of NO, it was demonstrated that valve endothelial-derived NO reduced matrix calcification in pro-osteogenic conditions, through soluble guanylyl cyclase binding and activation of cGMP signaling in VICs. Applying side-specific waveforms to porcine aortic valves, the authors demonstrated higher endothelial cGMP expression on the ventricularis side (high-magnitude shear stress). The addition of a soluble guanylyl cyclase inhibitor to block the pathway yielded an increase in α-SMA expression and aorta-side osteocalcin. This suggests that protective NO production by VECs is shear stress regulated and that side-specific hemodynamic shear stress profiles differentially regulate NO production to yield a differential protective effect.

The Notch1 pathway is implicated in disease protection and is likely regulated by shear stress. Much of this work has been studied in VIC models. In 2005, Garg et al. uncovered a heterozygous mutation in the Notch1 receptor gene, *NOTCH1* [83]. In vitro testing of a fibroblast cell line was used to test whether the expression of the Notch1 receptor directly affected osteogenesis. It was found that upon stimulation, the Notch1 intracellular cytoplasmic domain

(NICD) translocates to the nucleus to regulate gene signaling. Specifically, they found that Notch1-mediated repression of the Runx2 transcription factor led to the downregulation of the osteogenic genes osteocalcin and osteopontin. Interestingly, a more recent study suggested that the Notch1 pathway is regulated by NO [128]. Here, they showed that mouse lung endothelial cells attenuated spontaneous porcine VIC calcification in coculture and that this effect was abrogated with selective knockdown of eNOS. They then showed that L-NAME,[3] a known inhibitor of NO, led to an increase in VIC calcification, which was abrogated by transfection of the VICs with NICD (Fig. 3.3A). Together, these studies suggest that shear stress induction of eNOS/NO regulates Notch1-dependent attenuation of VIC osteogenesis.

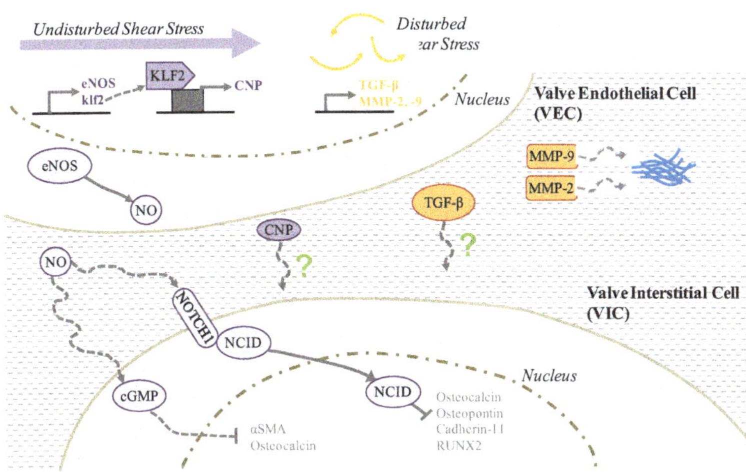

Figure 3.3 Emerging pathways involved in shear stress-mediated valve endothelial cell (VEC) paracrine signaling. (Light Purple) Undisturbed shear stress causes upregulation of the enzyme *eNOS* to synthesize nitric oxide (NO), secretion of which paracrine signals to neighboring valve interstitial cells (VICs) through cGMP and NOTCH1/NCID pathways to inhibit pathological gene expression. (Dark Purple) It is hypothesized that, as in the vasculature, shear stress upregulates expression of C-type natriuretic peptide (CNP), via the transcription factor *klf2*. CNP is protective against aortic valve disease, but the source and mechanism are not yet determined. (Orange) Conversely, disturbed shear stress causes upregulation of TGF-β and MMP-2 and MMP-9, all of which are higher expressed in disease aortic valves. However, the mechanisms remain to be elucidated.

[3] *N*-omega-nitro-L-arginine methyl ester.

Shear stress has also been shown to stimulate the Notch1 receptor and pathway in VECs directly. In one study, shear stress was shown to induce endothelial Notch1 cleavage and translocation to the nucleus, where it regulated the arterial endothelial cell marker ephrinB2 [129]. A more recent study showed that endothelial Notch1 mediates shear stress attenuation of calcification [130]. Using human iPSC-derived endothelial cells, the authors first demonstrated that shear stress antagonizes pro-osteogenic transforming growth factor-β (TGF-β), BMP4, and WNT pathways. However, this response to shear stress was not seen in Notch1 haploinsufficient (Notch$^{+/-}$) endothelial cells. These gene networks were further characterized to highlight epigenetic dysregulation in Notch1$^{+/-}$ endothelial cells. For example, Notch1 binding to the epigenetic marker H3K27ac was most repressed, which correlated with insufficient regulation of the anticalcific gene *ARHGEF17* and an increase in procalcific regulatory nodes SOX7 and TCF4. These novel studies have been comprehensive at uncovering shear stress–regulated landscapes at the transcriptional and epigenetic levels. They provide insight into how cells regulate homeostasis and reveal novel therapeutic targets for aortic valve disease intervention.

The downstream pathway of Notch1 signaling has also been associated with a decrease in cadherin-11, a transmembrane protein that forms cell-cell junctions with neighboring cells. Knockdown of cadherin-11 has been shown to inhibit dystrophic calcification in porcine VIC [131]. In a wound assay, cadherin-11 was shown to be involved in TGF-β1-induced cellular tension, suggesting this tension is necessary for calcific nodule development. Indeed, immunostaining of calcified human valves revealed that cadherin-11 expression was high in regions of significant calcification. Interestingly, Notch1$^{+/-}$ mouse VICs have been shown to have increased expression of cadherin-11 and decreased Runx2 activity, supporting the cadherin-11 implication in dystrophic calcification [132]. Functionally, overexpression of cadherin-11 in mouse hearts lead to Rho-A-dependent aortic valve calcification and ECM remodeling [133], and furthermore, blockage of cadherin-11 in Notch1-deficient mice abrogated the development of aortic valve disease [134]. It is not clear whether shear stress–regulated eNOS/NO/Notch1 signaling is implicated in both dystrophic and osteogenic calcification, but certainly this is a novel mechanistic insight into CAVD, with many possibilities for pharmacological intervention.

Lastly, Notch1 has been further implicated in BAV disease. A nonsense mutation in the *NOTCH1* gene has been shown to cause the BAV mutation [83, 135]. It is not known, however, whether BAVs are predisposed to CAVD due to downregulated Notch1 expression, whether Notch1 transmembrane receptor responds differently to BAV versus TAV waveforms, or whether the cumulative effect leads to accelerated disease in patients with BAVs.

3.4.3 Krüppel-Like Factor 2

KLF2 is a transcription factor predominantly expressed by endothelial cells exposed to shear stress. Well known as atheroprotective, KFL2 regulates pathways that mediate vascular tone [136] and vascular adhesion [137, 138]. In human endothelial cells, KLF2 expression has been shown to be induced by atheroprotective shear stress and not by atheroprone inflammatory cytokines, such as TNF-α[4] and IL-1β[5] [139]. In vascular endothelial cells exposed to atheroprotective or atherosusceptible waveforms of the vasculature, KLF2 was one of the most responsive genes upregulated by atheroprotective waveforms [140]. KLF2 functions through many pathways, including those involved in angiogenesis, inflammation, oxidative stress, and immune regulation [136]. To date, knowledge of the protective role of KLF2 signaling pathways in aortic valve homeostasis is incomplete.

In the aortic valve, KLF2 has been shown to be more highly expressed by VECs exposed to ventricular-side versus fibrosa-side shear stress waveforms [106, 112]. Downstream targets of KLF2 are the vasodilators NO [141] and CNP [138] (Fig. 3.3). As described above, CNP has been shown to be highly expressed on the disease-protected side of the aortic valve but not on the disease-prone side [109]. CNP has also been independently shown to be both shear stress regulated in the vasculature [94] and correlated with aortic valve disease prevention [142, 143]. Therefore, shear stress–induced KLF2 expression on the ventricular side of the aortic valve may maintain homeostasis via modulation of the CNP signaling pathway (Adams and Simmons, unpublished data, 2020).

[4]Tumor necrosis factor-α.

[5]Interleukin-1β.

3.4.4 Transforming Growth Factor-β

TGF-β is a potent cytokine that exerts its pleiotropic effects through canonical and noncanonical pathways [144]. The isoforms TGF-β1 and TGF-β3 are found in the vasculature and are known to play an imperative role in valve development and homeostasis [33], angiogenesis [145, 146], and apoptosis [147]. While TGF-β has long been implicated in pathological pathways, it has also been shown to be protective. For example, TGF-β1 induces myofibroblastic differentiation and pathological matrix remodeling in cultured VICs [32, 33], but conversely, TGF-β1 paracrine signaling from VECs has been shown to promote Sox9 nuclear localization in VICs, protecting against the development of calcific nodules [148]. It is not known whether local hemodynamics differentially regulate TGF-β expression levels to modulate its role in aortic valve disease progression.

TGF-β production and activation have been shown to be induced by shear stress in the vasculature. Platelets circulating in plasma secrete high levels of TGF-β1, which undergoes activation in response to an increase in shear stress levels [149]. In an aortic stenosis mouse model that used surgical aortic constriction to increase hemodynamic severity, ventricle-side shear stress levels were estimated to be as high as 1000 dyn/cm^2. Platelet-derived TGF-β1 was found on the aorta-side endothelium and associated with collagen-rich regions in the aortic valve tissue [150]. In a protective context, vascular endothelial cells have also been shown to express TGF-β3 in response to shear stress, mediating endothelial homeostasis through upregulation of KLF2 and NO [147], suggesting a similar role may exist in the aortic valve.

The role of TGF-β in valve homeostasis might be modulated by variations in shear stress waveform profiles and may therefore be a part of an adaptive response to valve pathogenesis. Porcine aortic valve leaflets have been studied in a novel ex vivo experimental setup that applied aorta-side and ventricle-side shear stress waveforms on their respective sides of tissue, as well as the opposite sides of the tissue. VEC inflammatory markers VCAM-1 and ICAM-1, as well as TGF-β1 and BMP4, were found to be upregulated on the aorta side of the valve exposed to altered (ventricle-side) shear stress waveforms [108]. This group then tested the independent effects of shear stress magnitude and shear stress frequency on valve endothelium. They found that TGF-β and BMP4 were more highly regulated by changes in shear stress magnitude compared with changes in shear stress frequency [151]. To further probe the effect of altered shear stress, the authors compared the effect of physiological fluid shear stress waveforms with waveforms of supraphysiologic frequency or magnitude [104]. Under normal flow conditions, valve homeostasis was maintained, whereas in response to supraphysiological magnitude waveforms, VECs were found to express BMP4 and TGF-β1-dependent inflammatory markers, as well as markers for ECM degradation (matrix metalloproteinase [MMP]-2 and MMP-9) (Fig. 3.3). These data suggest that altered shear stress leads to disproportionate, or pathological, expression of TGF-β1 to induce signaling pathways associated with aortic valve disease.

Altered shear stress–induced TGF-β1 signaling also suggests a mechanism for the role that BAV shear stress waveforms play in mechanically induced disease initiation. In the same ex vivo approach mentioned above, TAV and BAV shear stress waveforms were applied to normal porcine aortic valves. The authors demonstrated that aorta-side BAV waveforms initiated a procalcific phenotype of the fibrosa side tissue: endothelial activation (ICAM-1 and VCAM-1), BMP4 and TGF-β1 signaling, ECM remodeling (MMP-2 and MMP-9), and an increase in osteocalcin protein [152]. This work supports that altered BAV hemodynamics may play a role in CAVD progression, potentially making BAVs more vulnerable to disease at an earlier stage in life and to a greater extent.

3.5 Conclusions

The aortic valve is a unique component of the heart that is exposed to a series of mechanical forces throughout the cardiac cycle. Shear stress, the primary force acting on the valve, is caused by the flow of pressurized blood along the surface of the endothelium. The side-specific shear stress profiles of aortic valves are extensively studied using novel computational and experimental methods and have demonstrated that ventricle-side VECs experience high unidirectional shear stress, whereas aorta-side VECs are exposed to low, oscillatory shear stress. VECs sense shear stress and undergo phenotypic changes that can lead to the progression of pathological or protective signaling pathways, and it is widely hypothesized that aorta-side VECs are predisposed to disease progression as a result of the characteristics of their shear stress stimulus. Notably, phenotypes commonly associated with aortic valve disease, such as upregulation of VCAM-1 and TGF-β and induction of EndMT, are more common in the aorta-side endothelium as well as in VEC cultures exposed to aorta-side shear stress. Alternatively, protective mechanotransductive pathways related to NO, Notch1, and KLF2 have been shown to be more active on the ventricular-side endothelium. There is increasing evidence of, and appreciation for, the role of biomechanics in aortic valve biology; however, further work is required to explore how shear stress–induced VEC changes regulate VIC and aortic valve biology, in order to discover therapeutic interventions for aortic valve disease.

References

1. Freeman, R. V. (2005). Spectrum of calcific aortic valve disease: pathogenesis, disease progression, and treatment strategies. *Circulation*, **111**, pp. 3316–3326.
2. Stella, J. A., Liao, J. and Sacks, M. S. (2007). Time-dependent biaxial mechanical behavior of the aortic heart valve leaflet. *J Biomech*, **40**(14), pp. 3169–3177.
3. Chen, J.-H. and Simmons, C. A. (2011). Cell-matrix interactions in the pathobiology of calcific aortic valve disease: critical roles for matricellular, matricrine, and matrix mechanics cues. *Circ Res*, **108**, pp. 1510–1524.
4. Schoen, F. J. (2008). Evolving concepts of cardiac valve dynamics: the continuum of development, functional structure, pathobiology, and tissue engineering. *Circulation*, **118**, pp. 1864–1880.
5. Tseng, H., et al. (2013). The tensile and viscoelastic properties of aortic valve leaflets treated with a hyaluronidase gradient. *Cardiovasc Eng Technol*, **4**, pp. 151–160.
6. Li, C., Xu, S. and Gotlieb, A. I. (2013). The progression of calcific aortic valve disease through injury, cell dysfunction, and disruptive biologic and physical force feedback loops. *Cardiovasc Pathol*, **22**, pp. 1–8.
7. Ward, C. (2000). Clinical significance of the bicuspid aortic valve. *Heart*, **83**, pp. 81–85.

8. Stephan, P. J., Henry, A. C., Hebeler, R. F., Whiddon, L. and Roberts, W. C. (1997). Comparison of age, gender, number of aortic valve cusps, concomitant coronary artery bypass grafting, and magnitude of left ventricular-systemic arterial peak systolic gradient in adults having aortic valve replacement for isolated aortic valve stenosis. *Am J Cardiol*, **79**, pp. 166–72.

9. Laforest, B. and Nemer, M. (2012). Genetic insights into bicuspid aortic valve formation. *Cardiol Res Pract*, **2012**, pp. 1–8.

10. Sievers, H.-H. and Schmidtke, C. (2007). A classification system for the bicuspid aortic valve from 304 surgical specimens. *J Thorac Cardiovasc Surg*, **133**, pp. 1226–1233.

11. Yazdani Sabet, H., Edwards, W. D., Tazelaar, H. D. and Daly, R. C. (1999). Congenitally bicuspid aortic valves: a surgical pathology study of 542 cases (1991 through 1996) and a literature review of 2,715 additional cases. *Mayo Clin Proc*, **74**, pp. 14–26.

12. Atkins, S. K. (2014). Etiology of bicuspid aortic valve disease: focus on hemodynamics. *World J Cardiol*, **6**, pp. 1227.

13. Taylor, P. M., Allen, S. P. and Yacoub, M. H. (2000). Phenotypic and functional characterization of interstitial cells from human heart valves, pericardium and skin. *J Heart Valve Dis*, **9**, pp. 150–158.

14. Chen, J.-H., Chen, W. L. K., Sider, K. L., Yip, C. Y. Y. and Simmons, C. A. (2011). B-catenin mediates mechanically regulated, transforming growth factor- 1-induced myofibroblast differentiation of aortic valve interstitial cells. *Arterioscler Thromb Vasc Biol*, **31**, pp. 590–597.

15. Chen, J.-H., Yip, C. Y. Y., Sone, E. D. and Simmons, C. A. (2009). Identification and characterization of aortic valve mesenchymal progenitor cells with robust osteogenic calcification potential. *Am J Pathol*, **174**, pp. 1109–1119.

16. Tompkins, R. G., Schnitzer, J. J. and Yarmush, M. L. (1989). Macromolecular transport within heart valves. *Circ Res*, **64**, pp. 1213–1223.

17. Otto, C. M., Kuusisto, J., Reichenbach, D. D., Gown, A. M. and O'Brien, K. D. (1994). Characterization of the early lesion of 'degenerative' valvular aortic stenosis. Histological and immunohistochemical studies. *Circulation*, **90**, pp. 844–853.

18. Muller, A. M., et al. (2000). Expression of endothelial cell adhesion molecules on heart valves: up-regulation in degeneration as well as acute endocarditis. *J Pathol*, **191**, pp. 54–60.

19. Davies, P. F. (1997). Mechanisms involved in endothelial responses to hemodynamic forces. *Atherosclerosis*, **131**, pp. S15–S17.

20. Butcher, J. T. and Nerem, R. M. (2006). Valvular endothelial cells regulate the phenotype of interstitial cells in co-culture: effects of steady shear stress. *Tissue Eng*, **12**, pp. 905–915.

21. Chen, M. B., Srigunapalan, S., Wheeler, A. R. and Simmons, C. A. (2013). A 3D microfluidic platform incorporating methacrylated gelatin hydrogels to study physiological cardiovascular cell–cell interactions. *Lab Chip*, **13**, pp. 2591.

22. Aggarwal, A., et al. (2006). Architectural trends in the human normal and bicuspid aortic valve leaflet and its relevance to valve disease. *Ann Biomed Eng*, **42**(5), pp. 986–998.

23. Otto, C. M., et al. (1997). Prospective study of asymptomatic valvular aortic stenosis. Clinical, echocardiographic, and exercise predictors of outcome. *Circulation*, **95**, pp. 2262–2270.

24. Lindman, B. R., Bonow, R. O. and Otto, C. M. (2013). Current management of calcific aortic stenosis. *Circ Res*, **113**, pp. 223–237.

25. Hsu, S.-Y., Hsieh, I.-C., Chang, S.-H., Wen, M.-S. and Hung, K.-C. (2005). Aortic valve sclerosis is an echocardiographic indicator of significant coronary disease in patients undergoing diagnostic coronary angiography. *Int J Clin Pract*, **59**, pp. 72–77.

26. Otto, C. M. (1999). Doppler echocardiographic evaluation of aortic and mitral stenoses. *Echocardiography*, **16**, pp. 675–676.

27. Alwan, A. (2011). *Global Status Report on Noncommunicable Diseases 2010* (World Health Organization).

28. Messé, S. R., et al. (2014). Stroke after aortic valve surgery: results from a prospective cohort. *Circulation*, **129**, pp. 2253–2261.

29. Rajamannan, N. M., et al. (2011). Calcific aortic valve disease: not simply a degenerative process: a review and agenda for research from the National Heart and Lung and Blood Institute Aortic Stenosis Working Group * Executive summary: calcific aortic valve disease - 2011 update. *Circulation*, **124**, pp. 1783–1791.

30. Ghaisas, N. K., et al. (2000). Adhesion molecules in nonrheumatic aortic valve disease: endothelial expression, serum levels and effects of valve replacement. *J Am Coll Cardiol*, **36**, pp. 2257–2262.

31. Sider, K. L., et al. (2014). Evaluation of a porcine model of early aortic valve sclerosis. *Cardiovasc. Pathol*, **23**, pp. 289–297.

32. Walker, G. A., Masters, K. S., Shah, D. N., Anseth, K. S. and Leinwand, L. A. (2004). Valvular myofibroblast activation by transforming growth

factor-β: implications for pathological extracellular matrix remodeling in heart valve disease. *Circ Res*, **95**, pp. 253–260.

33. Cushing, M. C., Liao, J. T. and Anseth, K. S. (2005). Activation of valvular interstitial cells is mediated by transforming growth factor-B1 interactions with matrix molecules. *Matrix Biol*, **24**, pp. 428–437.
34. Hinz, B., Celetta, G., Tomasek, J. J., Gabbiani, G. and Chaponnier, C. (2001). Alpha-smooth muscle actin expression upregulates fibroblast contractile activity. *Mol Biol Cell*, **12**, pp. 2730–2741.
35. Mohler, E. R., et al. (2001). Bone formation and inflammation in cardiac valves. *Circulation*, **103**, pp. 1522–1528.
36. Roberts, W. C. (1970). The congenitally bicuspid aortic valve. A study of 85 autopsy cases. *Am J Cardiol*, **26**, pp. 72–83.
37. Stewart, B. F., et al. (1997). Clinical factors associated with calcific aortic valve disease fn1fn1This study was supported in part by contracts NO1-HC85079 through HC-850086 from the National Heart, Lung, and Blood Institute, National Institutes of Health, Bethesda, Maryland. *J Am Coll Cardiol*, **29**, pp. 630–634.
38. Mozaffarian, D., et al. (2015). Heart disease and stroke statistics—2015 update. *Circulation*, **131**, pp. e29–e322.
39. Siu, S. C. and Silversides, C. K. (2010). Bicuspid aortic valve disease. *J Am Coll Cardiol*, **55**(25), pp. 2789–2800.
40. Gould, S. T., Srigunapalan, S., Simmons, C. A. and Anseth, K. S. (2013). Hemodynamic and cellular response feedback in calcific aortic valve disease. *Circ Res*, **113**, pp. 186–197.
41. Davies, P. F. (1995). Flow-mediated endothelial mechanotransduction. *Physiol Rev*, **75**, pp. 519–560.
42. Conway, D. E., Williams, M. R., Eskin, S. G. and McIntire, L. V. (2010). Endothelial cell responses to atheroprone flow are driven by two separate flow components: low time-average shear stress and fluid flow reversal. *Am J Physiol Heart Circ Physiol*, **298**, pp. H367–H374.
43. Nanda, N. C., et al. (1974). Echocardiographic recognition of the congenital bicuspid aortic valve. *Circulation*, **49**, pp. 870–875.
44. Fowles, R. E., et al. (1979). Two-dimensional echocardiographic features of bicuspid aortic valve. *Chest*, **75**, pp. 434–440.
45. Barker, A. J., Lanning, C. and Shandas, R. (2010). Quantification of hemodynamic wall shear stress in patients with bicuspid aortic valve using phase-contrast MRI. *Ann Biomed Eng*, **38**, pp. 788–800.

46. Hope, M. D., et al. (2010). Bicuspid aortic valve: four-dimensional mr evaluation of ascending aortic systolic flow patterns. *Radiology*, **255**, pp. 53–61.

47. Osinnski, J. N., Ku, D. N., Mukundan, S., Loth, F. and Pettigrew, R. I. (1995). Determination of wall shear stress in the aorta with the use of MR phase velocity mapping. *J Magn Reson Imaging*, **5**, pp. 640–647.

48. Ortlepp, J. R. (2003). Cardiovascular risk factors in patients with aortic stenosis predict prevalence of coronary artery disease but not of aortic stenosis: an angiographic pair matched case-control study. *Heart*, **89**, pp. 1019–1022.

49. Farivar, R. S., et al. (2003). Transcriptional profiling and growth kinetics of endothelium reveals differences between cells derived from porcine aorta versus aortic valve. *Eur J Cardio-thoracic Surg*, **24**, pp. 527–534.

50. Young, E. W. K., Wheeler, A. R. and Simmons, C. A. (2007). Matrix-dependent adhesion of vascular and valvular endothelial cells in microfluidic channels. *Lab Chip*, **7**, pp. 1759.

51. Butcher, J. T. and Nerem, R. M. (2007). Valvular endothelial cells and the mechanoregulation of valvular pathology. *Philos Trans R Soc B Biol Sci*, **362**, pp. 1445–1457.

52. Robicsek, F., Thubrikar, M. J., Cook, J. W. and Fowler, B. (2004). The congenitally bicuspid aortic valve: how does it function? Why does it fail? *Ann Thorac Surg*, **77**, pp. 177–185.

53. Walburn, F. J. and Stein, P. D. (1984). Wall shear stress during pulsatile flow distal to a normal porcine aortic valve. *J Biomech*, **17**, pp. 97–102.

54. Weston, M. W., LaBorde, D. V and Yoganathan, A. P. (1999). Estimation of the shear stress on the surface of an aortic valve leaflet. *Ann Biomed Eng*, **27**, pp. 572–579.

55. Saikrishnan, N., Mirabella, L. and Yoganathan, A. P. (2015). Bicuspid aortic valves are associated with increased wall and turbulence shear stress levels compared to trileaflet aortic valves. *Biomech Model Mechanobiol*, **14**, pp. 577–588.

56. Saikrishnan, N., Yap, C.-H., Milligan, N. C., Vasilyev, N. V and Yoganathan, A. P. (2012). In vitro characterization of bicuspid aortic valve hemodynamics using particle image velocimetry. *Ann Biomed Eng*, **40**, pp. 1760–1775.

57. Markl, M., Frydrychowicz, A., Kozerke, S., Hope, M. and Wieben, O. (2012). 4D flow MRI. *J Magn Reson Imaging*, **36**, pp. 1015–1036.

58. Markl, M., Schnell, S. and Barker, A. J. (2014). 4D flow imaging: current status to future clinical applications. *Curr Cardiol Rep*, **16**, pp. 481.

59. De Hart, J., Peters, G. W. M., Schreurs, P. J. G. and Baaijens, F. P. T. (2003). A three-dimensional computational analysis of fluid–structure interaction in the aortic valve. *J Biomech*, **36**, pp. 103–112.

60. Chandra, S., Rajamannan, N. M. and Sucosky, P. (2012). Computational assessment of bicuspid aortic valve wall-shear stress: implications for calcific aortic valve disease. *Biomech Model Mechanobiol*, **11**, pp. 1085–1096.

61. Sotelo, J., et al. (2018). 3D axial and circumferential wall shear stress from 4D flow MRI data using a finite element method and a laplacian approach. *Magn Reson Med*, **79**, pp. 2816–2823.

62. Piatti, F., et al. (2017). Towards the improved quantification of in vivo abnormal wall shear stresses in BAV-affected patients from 4D-flow imaging: Benchmarking and application to real data. *J Biomech*, **50**, pp. 93–101.

63. Carmody, C. J., Burriesci, G., Howard, I. C. and Patterson, E. A. (2006). An approach to the simulation of fluid–structure interaction in the aortic valve. *J Biomech*, **39**, pp. 158–169.

64. Morsi, Y. S., Yang, W. W., Wong, C. S. and Das, S. (2007). Transient fluid–structure coupling for simulation of a trileaflet heart valve using weak coupling. *J Artif Organs*, **10**, pp. 96–103.

65. Yap, C. H., Saikrishnan, N., Tamilselvan, G., Vasilyev, N. and Yoganathan, A. P. (2012). The congenital bicuspid aortic valve can experience high-frequency unsteady shear stresses on its leaflet surface. *AJP Hear Circ Physiol*, **303**, pp. H721–H731.

66. Ge, L. and Sotiropoulos, F. (2010). Direction and magnitude of blood flow shear stresses on the leaflets of aortic valves: is there a link with valve calcification? *J Biomech Eng*, **132**, pp. 014505.

67. Yap, C. H., Saikrishnan, N., Tamilselvan, G. and Yoganathan, A. P. (2012). Experimental measurement of dynamic fluid shear stress on the aortic surface of the aortic valve leaflet. *Biomech Model Mechanobiol*, **11**, pp. 171–182.

68. Weinberg, E. J. and Kaazempur Mofrad, M. R. (2008). A multiscale computational comparison of the bicuspid and tricuspid aortic valves in relation to calcific aortic stenosis. *J Biomech*, **41**, pp. 3482–3487.

69. Bellhouse, B. J. and Talbot, L. (1969). The fluid mechanics of the aortic valve. *J Fluid Mech*, **35**, pp. 721.

70. Markl, M., et al. (2005). Time-resolved three-dimensional magnetic resonance velocity mapping of aortic flow in healthy volunteers

and patients after valve-sparing aortic root replacement. *J Thorac Cardiovasc Surg*, **130**, pp. 456–463.

71. Cao, K., Bukac, M. and Sucosky, P. (2016). Three-dimensional macro-scale assessment of regional and temporal wall shear stress characteristics on aortic valve leaflets. *Comput. Methods Biomech Biomed Eng*, **19**, pp. 603–613.

72. Cao, K., Atkins, S. K., McNally, A., Liu, J. and Sucosky, P. (2017). Simulations of morphotype-dependent hemodynamics in non-dilated bicuspid aortic valve aortas. *J Biomech*, **50**, pp. 63–70.

73. Nkomo, V. T., et al. (2003). Bicuspid aortic valve associated with aortic dilatation: a community-based study. *Arterioscler Thromb Vasc Biol*, **23**, pp. 351–356.

74. Bauer, M., Siniawski, H., Pasic, M., Schaumann, B. and Hetzer, R. (2006). Different hemodynamic stress of the ascending aorta wall in patients with bicuspid and tricuspid aortic valve. *J Card Surg*, **21**, pp. 218–220.

75. Bissell, M. M., et al. (2013). Aortic dilation in bicuspid aortic valve disease: flow pattern is a major contributor and differs with valve fusion type. *Circ Cardiovasc Imaging*, **6**, pp. 499–507.

76. Atkins, S. K., Cao, K., Rajamannan, N. M. and Sucosky, P. (2014). Bicuspid aortic valve hemodynamics induces abnormal medial remodeling in the convexity of porcine ascending aortas. *Biomech Model Mechanobiol*, **13**, pp. 1209–1225.

77. Hope, M. D., Sigovan, M., Wrenn, S. J., Saloner, D. and Dyverfeldt, P. (2014). MRI hemodynamic markers of progressive bicuspid aortic valve-related aortic disease. *J Magn Reson Imaging*, **40**, pp. 140–145.

78. Guzzardi, D. G., et al. (2015). Valve-related hemodynamics mediate human bicuspid aortopathy. *J Am Coll Cardiol*, **66**, pp. 892–900.

79. van Ooij, P., et al. (2015). Characterization of abnormal wall shear stress using 4D flow MRI in human bicuspid aortopathy. *Ann Biomed Eng*, **43**, pp. 1385–1397.

80. Conti, C. A., et al. (2010). Biomechanical implications of the congenital bicuspid aortic valve: a finite element study of aortic root function from in vivo data. *J Thorac Cardiovasc Surg*, **140**, pp. 890–896.e2.

81. Lorenz, R., et al. (2014). 4D flow magnetic resonance imaging in bicuspid aortic valve disease demonstrates altered distribution of aortic blood flow helicity. *Magn Reson Med*, **71**, pp. 1542–1553.

82. Lee, T. C., Zhao, Y. D., Courtman, D. W. and Stewart, D. J. (2000). Abnormal aortic valve development in mice lacking endothelial nitric oxide synthase. *Circulation*, **101**, pp. 2345–2348.

83. Garg, V., et al. (2005). Mutations in NOTCH1 cause aortic valve disease. *Nature*, **437**, pp. 270–274.

84. Laforest, B., Andelfinger, G. and Nemer, M. (2011). Loss of Gata5 in mice leads to bicuspid aortic valve. *J Clin Invest*, **121**(7), pp. 2876–2887.

85. Marom, G., Kim, H.-S., Rosenfeld, M., Raanani, E. and Haj-Ali, R. (2013). Fully coupled fluid–structure interaction model of congenital bicuspid aortic valves: effect of asymmetry on hemodynamics. *Med Biol Eng Comput*, **51**, pp. 839–848.

86. Feaver, R. E., Gelfand, B. D. and Blackman, B. R. (2013). Human haemodynamic frequency harmonics regulate the inflammatory phenotype of vascular endothelial cells. *Nat Commun*, **4**, pp. 1525.

87. Ni, C. W., et al. (2010). Discovery of novel mechanosensitive genes in vivo using mouse carotid artery endothelium exposed to disturbed flow. *Blood*, **116**, pp. e66–e73.

88. Thijssen, D. H. J., Dawson, E. A., Tinken, T. M., Cable, N. T. and Green, D. J. (2009). Retrograde flow and shear rate acutely impair endothelial function in humans. *Hypertension*, **53**, pp. 986–992.

89. Papadaki, M. and Eskin, S. G. (1997). Effects of fluid shear stress on gene regulation of vascular cells. *Biotechnol Prog*, **13**, pp. 209–221.

90. Humphrey, J. D., Dufresne, E. R. and Schwartz, M. A. (2014). Mechanotransduction and extracellular matrix homeostasis. *Nat Rev Mol Cell Biol*, **15**(12), pp. 802–812.

91. Zhou, J., Li, Y.-S. and Chien, S. (2014). Shear stress-initiated signaling and its regulation of endothelial function. *Arterioscler Thromb Vasc Biol*, **34**, pp. 2191–2198.

92. Buga, G. M., Gold, M. E., Fukuto, J. M. and Ignarro, L. J. (1991). Shear stress-induced release of nitric oxide from endothelial cells grown on beads. *Hypertension*, **17**, pp. 187–193.

93. Ranjan, V., Xiao, Z. and Diamond, S. L. (1995). Constitutive NOS expression in cultured endothelial cells is elevated by fluid shear stress. *Am J Physiol Circ Physiol*, **269**, pp. H550–H555.

94. Zhang, Z., Xiao, Z. and Diamond, S. L. (1999). Shear stress induction of C-type natriuretic peptide (CNP) in endothelial cells is independent of NO autocrine signaling. *Ann Biomed Eng*, **27**, pp. 419–426.

95. Hwang, J., et al. (2003). Oscillatory shear stress stimulates endothelial production of from p47 phox -dependent NAD(P)H oxidases, leading to monocyte adhesion. *J Biol Chem*, **278**, pp. 47291–47298.

96. Topper, J. N., Cai, J., Falb, D. and Gimbrone, M. A. (1996). Identification of vascular endothelial genes differentially responsive to fluid mechanical

stimuli: cyclooxygenase-2, manganese superoxide dismutase, and endothelial cell nitric oxide synthase are selectively up-regulated by steady laminar shear stress. *Proc Natl Acad Sci U S A*, **93**, pp. 10417–10422.

97. Noris, M., et al. (1995). Nitric oxide synthesis by cultured endothelial cells is modulated by flow conditions. *Circ Res*, **76**, pp. 536–543.

98. Garcia-Cardeña, G., Comander, J., Anderson, K. R., Blackman, B. R. and Gimbrone, M. A. (2001). Biomechanical activation of vascular endothelium as a determinant of its functional phenotype. *Proc Natl Acad Sci U S A*, **98**, pp. 4478–4485.

99. Fernández Esmerats, J., Heath, J. and Jo, H. (2016). Shear-sensitive genes in aortic valve endothelium. *Antioxid Redox Signal*, **25**, pp. 401–414.

100. Srigunapalan, S., Lam, C., Wheeler, A. R. and Simmons, C. A. (2011). A microfluidic membrane device to mimic critical components of the vascular microenvironment. *Biomicrofluidics*, **5**(1), pp. 013409.

101. Young, E. W. K., Watson, M. W. L., Srigunapalan, S., Wheeler, A. R. and Simmons, C. A. (2010). Technique for real-time measurements of endothelial permeability in a microfluidic membrane chip using laser-induced fluorescence detection. *Anal Chem*, **82**, pp. 808–816.

102. Blackman, B. R. and Gimbrone, M. A. (2002). A new in vitro model to evaluate differential responses of endothelial cells to simulated. *Image (Rochester, NY)*, **124**, pp. 397–407.

103. Sucosky, P., et al. (2008). Design of an ex vivo culture system to investigate the effects of shear stress on cardiovascular tissue. *J Biomech Eng*, **130**, pp. 035001.

104. Sun, L. and Sucosky, P. (2015). Bone morphogenetic protein-4 and transforming growth factor-beta1 mechanisms in acute valvular response to supra-physiologic hemodynamic stresses. *World J Cardiol*, **7**, pp. 331–343.

105. Butcher, J. T., Penrod, A. M., García, A. J. and Nerem, R. M. (2004). Unique morphology and focal adhesion development of valvular endothelial cells in static and fluid flow environments. *Arterioscler Thromb Vasc Biol*, **24**, pp. 1429–1434.

106. Holliday, C. J., Ankeny, R. F., Jo, H. and Nerem, R. M. (2011). Discovery of shear- and side-specific mRNAs and miRNAs in human aortic valvular endothelial cells. *AJP Hear Circ Physiol*, **301**, pp. H856–H867.

107. Deck, J. D. (1986). Endothelial cell orientation on aortic valve leaflets. *Cardiovasc Res*, **20**, pp. 760–767.

108. Sucosky, P., Balachandran, K., Elhammali, A., Jo, H. and Yoganathan, A. P. (2009). Altered shear stress stimulates upregulation of endothelial VCAM-1 and ICAM-1 in a BMP-4- and TGF-beta1-dependent pathway. *Arterioscler Thromb Vasc Biol*, **29**, pp. 254–260.

109. Simmons, C. A. (2005). Spatial heterogeneity of endothelial phenotypes correlates with side-specific vulnerability to calcification in normal porcine aortic valves. *Circ Res*, **96**, pp 792–799.

110. Butcher, J. T., et al. (2006). Transcriptional profiles of valvular and vascular endothelial cells reveal phenotypic differences: influence of shear stress. *Arterioscler Thromb Vasc Biol*, **26**(1), pp. 69–77.

111. Xia, M. and Sui, Z. (2009). Recent developments in CCR2 antagonists. *Expert Opin Ther Pat*, **19**, pp. 295–303.

112. Weinberg, E. J., Mack, P. J., Schoen, F. J., García-Cardeña, G. and Kaazempur Mofrad, M. R. (2010). Hemodynamic environments from opposing sides of human aortic valve leaflets evoke distinct endothelial phenotypes in vitro. *Cardiovasc Eng*, **10**, pp. 5–11.

113. Heath, J. M., et al. (2017). Mechanosensitive microRNA-181b regulates aortic valve endothelial matrix degradation by targeting TIMP3. *Cardiovasc Eng Technol*, **9**, pp. 141–150.

114. Miragoli, M., et al. (2014). Side-specific mechanical properties of valve endothelial cells. *AJP Hear Circ Physiol*, **307**, pp. H15–H24.

115. Markwald, R. R., Fitzharris, T. P. and Manasek, F. J. (1997). Structural development of endocardial cushions. *Am J Anat*, **148**, pp. 85–119.

116. Paranya, G., et al. (2001). Aortic valve endothelial cells undergo transforming growth factor-β-mediated and non-transforming growth factor-β-mediated transdifferentiation in vitro. *Am J Pathol*, **159**, pp. 1335–1343.

117. Mahler, G. J., Frendl, C. M., Cao, Q. and Butcher, J. T. (2014). Effects of shear stress pattern and magnitude on mesenchymal transformation and invasion of aortic valve endothelial cells. *Biotechnol Bioeng*, **111**(11), pp. 2326–2337.

118. Balachandran, K., et al. (2011). Cyclic strain induces dual-mode endothelial-mesenchymal transformation of the cardiac valve. *Proc Natl Acad Sci U S A*, **108**, pp. 19943–19948.

119. Zhong, A., Mirzaei, Z. and Simmons, C. A. (2018). The roles of matrix stiffness and ß-catenin signaling in endothelial-to-mesenchymal transition of aortic valve endothelial cells. *Cardiovasc Eng Technol*, **9**, pp. 158–167.

120. Palmer, R. M. J., Ashton, D. S. and Moncada, S. (1988). Vascular endothelial cells synthesize nitric oxide from L-arginine. *Nature*, **333**, pp. 664–666.

121. Huang, P. L., et al. (1995). Hypertension in mice lacking the gene for endothelial nitric oxide synthase. *Nature*, **377**, pp. 239–242.

122. Rudic, R. D., et al. (1998). Direct evidence for the importance of endothelium-derived nitric oxide in vascular remodeling. *J Clin Invest*, **101**, pp. 731–736.

123. Moncada, S. and Higgs, A. (1993). The L-arginine-nitric oxide pathway. *N Engl J Med*, **329**, pp. 2002–2012.

124. Kennedy, J. A., et al. (2009). Inhibition of calcifying nodule formation in cultured porcine aortic valve cells by nitric oxide donors. *Eur J Pharmacol*, **602**, pp. 28–35.

125. El Accaoui, R. N., et al. (2014). Aortic valve sclerosis in mice deficient in endothelial nitric oxide synthase. *AJP Hear Circ Physiol*, **306**, pp. H1302–H1313.

126. Uematsu, M., et al. (1995). Regulation of endothelial cell nitric oxide synthase mRNA expression by shear stress. *Am J Physiol Heart Circ Physiol*, **269**, pp. C1371– C1378.

127. Richards, J., et al. (2013). Side-specific endothelial-dependent regulation of aortic valve calcification: interplay of hemodynamics and nitric oxide signaling. *Am J Pathol*, **182**(5), pp. 1922–1931.

128. Bosse, K., et al. (2013). Endothelial nitric oxide signaling regulates Notch1 in aortic valve disease. *J Mol Cell Cardiol*, **60**, pp. 27–35.

129. Masumura, T., Yamamoto, K., Shimizu, N., Obi, S. and Ando, J. (2009). Shear stress increases expression of the arterial endothelial marker ephrinB2 in murine ES cells via the VEGF-notch signaling pathways. *Arterioscler Thromb Vasc Biol*, **29**, pp. 2125–2131.

130. Theodoris, C. V., et al. (2015). Human disease modeling reveals integrated transcriptional and epigenetic mechanisms of NOTCH1 haploinsufficiency. *Cell*, **160**, pp. 1072–1086.

131. Hutcheson, J. D., et al. (2013). Cadherin-11 regulates cell-cell tension necessary for calcific nodule formation by valvular myofibroblasts. *Arterioscler Thromb Vasc Biol*, **33**, pp. 114–120.

132. Chen, J., et al. (2015). Notch1 mutation leads to valvular calcification through enhanced myofibroblast mechanotransduction significance. *Arterioscler Thromb Vasc Biol*, **35**, pp. 1597–1605.

133. Sung, D. C., et al. (2016). Cadherin-11 overexpression induces extracellular matrix remodeling and calcification in mature aortic valves highlights. *Arterioscler Thromb Vasc Biol*, **36**, pp. 1627–1637.

134. Clark, C. R., Bowler, M. A., Snider, J. C. and Merryman, W. D. (2017). Targeting cadherin-11 prevents notch1-mediated calcific aortic valve disease. *Circulation*, **135**, pp. 2448–2450.

135. McKellar, S. H., et al. (2007). Novel NOTCH1 mutations in patients with bicuspid aortic valve disease and thoracic aortic aneurysms. *J Thorac Cardiovasc Surg*, **134**, pp. 290–296.

136. Nayak, L., Lin, Z. and Jain, M. K. (2011). 'Go with the flow': how Krüppel-like factor 2 regulates the vasoprotective effects of shear stress. *Antioxid Redox Signal*, **15**, pp. 1449–1461.

137. SenBanerjee, S., et al. (2004). KLF2 is a novel transcriptional regulator of endothelial proinflammatory activation. *J Exp Med*, **199**, pp. 1305–1315.

138. Parmar, K. M., et al. (2006). Integration of flow-dependent endothelial phenotypes by Kruppel-like factor 2. *J Clin Invest*, **116**, pp. 49–58.

139. Dekker, R. J., et al. (2002). Prolonged fluid shear stress induces a distinct set of endothelial cell genes, most specifically lung Krüppel-like factor (KLF2). *Blood*, **100**, pp. 1689–1698.

140. Dai, G., et al. (2004). Distinct endothelial phenotypes evoked by arterial waveforms derived from atherosclerosis-susceptible and -resistant regions of human vasculature. *Proc Natl Acad Sci U S A*, **101**, pp. 14871–14876.

141. Slater, S. C., et al. (2012). Chronic exposure to laminar shear stress induces Kruppel-like factor 2 in glomerular endothelial cells and modulates interactions with co-cultured podocytes. *Int J Biochem Cell Biol*, **44**, pp. 1482–1490.

142. Yip, C. Y. Y., Blaser, M. C., Mirzaei, Z., Zhong, X. and Simmons, C. A. (2011). Inhibition of pathological differentiation of valvular interstitial cells by C-type natriuretic peptide. *Arterioscler Thromb Vasc Biol*, **31**, pp. 1881–1889.

143. Blaser, M. C., et al. (2018). Deficiency of natriuretic peptide receptor 2 promotes bicuspid aortic valves, aortic valve disease, left ventricular dysfunction, and ascending aortic dilatations in mice. *Circ Res*, **122**, pp. 405–416.

144. Lawrence, D. A. (1996). Transforming growth factor-beta: a general review. *Eur Cytokine Netw*, **7**, pp. 363–374.

145. Jarad, M., Kuczynski, E. A., Morrison, J., Viloria-Petit, A. M. and Coomber, B. L. (2017). Release of endothelial cell associated VEGFR2 during TGF-β modulated angiogenesis in vitro. *BMC Cell Biol*, **18**, pp. 10.

146. Ferrari, G., Cook, B. D., Terushkin, V., Pintucci, G. and Mignatti, P. (2009). Transforming growth factor-beta 1 (TGF-β1) induces angiogenesis through vascular endothelial growth factor (VEGF)-mediated apoptosis. *J Cell Physiol*, **219**, pp. 449–458.

147. Walshe, T. E., Dela Paz, N. G. and D'Amore, P. A. (2013). The role of shear-induced transforming growth factor-β signaling in the endothelium. *Arterioscler Thromb Vasc Biol*, **33**, pp. 2608–2617.

148. Huk, D. J., et al. (2016). Valve endothelial cell–derived Tgfβ1 signaling promotes nuclear localization of Sox9 in interstitial cells associated with attenuated calcification significance. *Arterioscler Thromb Vasc Biol*, **36**, pp. 328–338.

149. Ahamed, J., et al. (2008). In vitro and in vivo evidence for shear-induced activation of latent transforming growth factor-1. *Blood*, **112**, pp. 3650–3660.

150. Wang, W., Vootukuri, S., Meyer, A., Ahamed, J. and Coller, B. S. (2014). Association between shear stress and platelet-derived transforming growth factor-1 release and activation in animal models of aortic valve stenosis. *Arterioscler Thromb Vasc Biol*, **34**, pp. 1924–1932.

151. Sun, L., Rajamannan, N. M. and Sucosky, P. (2013). Defining the role of fluid shear stress in the expression of early signaling markers for calcific aortic valve disease. *PLoS One*, **8**, pp. e84433.

152. Sun, L., Chandra, S. and Sucosky, P. (2012). Ex vivo evidence for the contribution of hemodynamic shear stress abnormalities to the early pathogenesis of calcific bicuspid aortic valve disease. *PLoS One*, **7**, pp. e48843.

Chapter 4

Mechanotransduction of Cardiovascular Development and Regeneration

Quinton Smith,[a,b] Justin Lowenthal,[b,c,d] and Sharon Gerecht[a,b,c,e]

[a]*Chemical and Biomolecular Engineering, Johns Hopkins University, Baltimore, MD 21218, USA*
[b]*Institute for NanoBioTechnology, Johns Hopkins University, Baltimore, MD 21218, USA*
[c]*Biomedical Engineering, Johns Hopkins University School of Medicine, Baltimore, MD 21205, USA*
[d]*Medical Scientist Training Program, Johns Hopkins University School of Medicine, Baltimore, MD 21205, USA*
[e]*Materials Science and Engineering, Johns Hopkins University, Baltimore, MD 21218, USA*
gerecht@jhu.edu

Mechanistic insights into human cardiovascular development and regeneration have been largely focused on biochemical signaling pathways manipulated by chemokines and growth factors. However, it is becoming increasingly well illustrated that mechanical cues play important roles during development, maturation, and regeneration

Modern Mechanobiology: Convergence of Biomechanics, Development, and Genomics
Edited by Juhyun Lee, Sharon Gerecht, Hanjoong Jo, and Tzung Hsiai

ISBN 978-981-4800-58-7 (Hardcover), 978-0-429-29483-9 (eBook)
www.jennystanford.com

of the heart and vasculature. This chapter first discusses the changing mechanical context of the cardiovascular niche from an embryologic development perspective. In the following sections, engineering approaches to mimic physical cues present in vascular morphogenesis, angiogenesis, and cardiac development are explored. The final section portrays examples of the implementation of engineering platforms to study cardiovascular differentiation and its impact on downstream maturation events.

4.1 Introduction

The human cardiovascular and circulatory systems are engineering marvels, comprising a highly coordinated system requiring the synchrony of a four-chambered pump under precise electromechanical control and a highly branched, active, hormonally responsive network of vessels. This system facilitates the circulation of nearly 5–6 liters of blood per minute, enabling transport of nutrients, oxygen, endocrine signaling, immunologic responses, and waste throughout the body via a highly complex, hierarchical branched network of blood vessels. This system can be viewed and modeled in relatively simple pressure-flow-resistance relationships. In clinical terms, therapies for blood pressure and heart failure are often aimed at manipulating these relationships: adjusting heart rate, cardiac contractility, systemic resistance (afterload), or several of these at once.

In reality, however, the cardiovascular system behaves according to a much more complex network of mechanical, biophysical, hormonal, metabolic, and electrical inputs. The heart performs coordinated mechanical activity directed through an electrically active syncytium; the arterial vasculature responds to shear stress and signaling through baroreceptors and chemoreceptors, and central autonomic input provides high-level governance of pressure and flow. The entire system responds by listening to autocrine, paracrine, and endocrine ion exchange and hormonal signaling in response to both physiologic and pathologic cell- and tissue-level behaviors (cross talk between endothelium and smooth muscle in physiological states, myofibroblast differentiation and proliferation in the pathogenesis of fibrosis, etc.). A multitude of cell types are ultimately involved across multiple organs in this cross talk.

To understand the studies that focus on modeling and engineering this system, it is first necessary understand the structure and dynamics of cardiovascular physiology. Thus, this chapter begins with an overview of the anatomy, physiology, and developmental steps leading to mature function of the heart and vasculature. Specific emphasis is placed on cell sourcing and the advantages and drawbacks of using neonatal or pluripotent sources. Finally, tools to create new in vitro cardiovascular system models and ways to manipulate them in vivo using bioengineering interventions are discussed, with a focus on biological microelectromechanical systems (BioMEMS) technologies.

4.2 A Primer on Cardiovascular Anatomy and Physiology

4.2.1 Cardiovascular Anatomy

The myocardial, endocardial, and epicardial layers that comprise the heart act in concert to pump blood throughout two separate circulatory systems, systemic and pulmonary, via two separate "pumps" split by the interatrial and interventricular septa. In the systemic circulatory system, deoxygenated blood entering from the pulmonary veins passes through the left atrium, through the left atrioventricular (AV) valve, finally reaching the left ventricle, where it is pumped for reoxygenation. Blood leaves the aortic semilunar valve to the aortic arch and thoracoabdominal aortic segments, passing through a systemic network of arterial blood vessels. These branches of arterial networks steadily become smaller, with decreasing smooth muscle tone and increasing fenestration, eventually forming arterioles and capillaries where diffusional oxygen exchange occurs. Deoxygenated oxygenated blood returns to the superior and inferior vena cavae, as the small capillaries from the arterial system transition to high-endothelial venules, fuse into large venules, and then empty into veins such as the vena cavae and the pulmonary vein [1, 2].

In the pulmonary circulatory system, oxygen-depleted blood is pumped to the lungs via the pulmonary artery, where it is reoxygenated and returned to the heart via the pulmonary vein.

During development, the cardiovascular system emerges with several structurally distinct pathways and shunts that are not present in the mature adult; for example, the ductus arteriosus carries blood shunted from the fetal pulmonary artery to the descending aorta. In addition, the foramen ovale allows direct passage of blood from the right to left atrium in the fetal heart. As a result of the increasing demand for oxygenated blood to the developing fetus, maternal cardiac output and blood flow volume is substantially increased.

Specifically, fetal blood is oxygenated in umbilical vessels via maternal uterine arteries. The unique oxygen carrying properties of fetal blood are further enhanced through the activity of fetal hemoglobin. Upon initial breaths after a newborn emerges during birth, dramatic changes in the distribution of vascular resistance cause several of these shunts to close and eventually atrophy, and the foramen ovale normally seals due to increased left-sided heart pressures [3, 4].

4.2.2 Heart Development

Heart development proceeds from mesendodermal commitment and migration from the epiblast and trilaminar disc during gastrulation, through the formation of a single heart tube and the differentiation of multipotent second heart field tissue from that of the first heart field destined to contribute to the left ventricular myocardium [5]. Second heart field tissue, meanwhile, migrates in response to physiochemical gradients to make up myocardium, coronary endothelial, smooth muscle, and nodal tissue in the right ventricle, atria, and outflow tracts [6, 7]. Contributions to the endothelium of the heart and outflow tracts come through the second heart field and also through distinct mesendodermal contributions to the vasa vasorum of the great vessels [4].

Although both adult and fetal cardiac tissues comprise fibroblasts, blood vessels, and cardiomyocytes, the relative proportions of these constituents change through the adult life. During embryonic development, the ratio of proliferation between cardiomyocytes and fibroblasts is nearly the same until birth, where cardiomyocytes achieve growth via hypertrophy rather than hyperplasia, while fibroblasts continue to proliferate [8–12]. In a fully matured adult, only about a third of the resident cells in the heart are cardiomyocytes, even though they comprise about 75% of the tissues volume. Cardiac fibroblasts (or, more broadly, stromal cells of mesenchymal origin) are essential in the homeostasis of cardiac tissue, producing regulatory extracellular matrix (ECM) proteins that contribute to heart functionality by providing regulatory cues during cardiac development (aiding in action potential propagation) and contributing to disease states during pathogenesis [13–16].

Mature skeletal muscle is derived from a highly proliferative population of progenitor myoblasts, with proliferation under the stimulation of growth factors from the fibroblast growth factor (FGF) family [17, 18]. Cell growth is hampered when these proliferation cues are dampened, and they undergo transcriptional modification, where they are differentiated into multinucleated myotubes. Satellite cells are quiescent stem myoblast progenitor cells that do not undergo maturation to myotubes, and they aid in skeletal muscle regeneration in response to injury, serving as the only putative stem cell population in the heart. Adult skeletal muscle is derived from differentiated myotubes that steadily become mature and grow via hypertrophy. Experimental evidence indicates that electromechanical junctions exist between cardiomyocytes and skeletal muscle, through synchrony in calcium transients, indicative of gap junction coupling. In addition to cardiomyocytes and fibroblasts, the heart contains sympathetic and vagal nerve cells, which extend throughout the myocardium [17–21].

4.2.3 Vascular Development

About 4 weeks after conception, an endothelialized plexus facilitating the delivery of oxygen and nutrients has already been established, primed to support shear forces from the newly formed contracting heart [22]. The plasticity of the vasculature that develops is attuned to its specialized function, dependent upon the location within the body. As a result, there exists considerable variability in the structure and function of the endothelial, fibroblast, pericyte, and smooth muscle cell populations that make up the vasculature. Endothelial cells (ECs) arise after splanchnic mesoderm differentiation, similar to the rise of cells that will make up the endocardium. ECs emerge in a 2D fashion, until they coalesce to form3D tubular structures, coaxed by the ECM and signaling pathways involving the vascular endothelial growth factor (VEGF), FGF, and transforming growth factor (TGF-β). After de novo vasculogenesis, the established vasculature continues to branch and extend to the somatic mesoderm in a process termed "angiogenesis" [23].

Arterial and venous identity are established at this stage0 and can be demarked by ephrinB2, DLL4, NRP-1, and Notch3 for arterial cells and Ephb4, COUPTFII, and Nrp2 for venous identity [24]. Although specification is established at this point, EC identity is plastic and can be pruned on the basis of the blood flow [25]. ECs that appear as cuboidal squamous cells in veins are phenotypically aligned in arterial vessels in response to the shear demands in systemic circulation. In addition, ECs in capillary beds take on varying phenotypes depending on tissue specialization. For example, in the bone marrow, liver, and spleen, capillary ECs are sinusoidal, permitting the direct contact of blood with underlying tissue for metabolic exchange [26]. On the other hand, in endocrine and exocrine glands, ECs are fenestrated to allow the exchange of hormones [27].

Subsequent to vasculogenesis, the endothelial scaffold is stabilized by supporting mural cell populations. Smooth muscle cells are recruited to larger arteries and veins, while pericytes associate with small-diameter capillaries [28]. Although the majority of smooth muscle cells arise from the mesoderm, smooth muscle cells also arise from ectodermal cardiac neural crest cells specifically to vasculature in the head and thoracic regions of the body. In addition, smooth muscle cells that make up coronary vessels stem from the proepicardium. The stabilization of these vessel structures is coaxed by the secretion of platelet-derived growth factor (PDGF-BB) by ECs to platelet-derived growth factor receptor 1 (PDGFR-1)-expressing mural and stromal cell populations. Subsequent to this signal, smooth muscle cells and pericytes secrete angiopoietins, which bind to Tie receptors on ECs, reducing their angiogenic activity [29].

4.3 Mechanics of the Cardiovascular System

4.3.1 Cardiac Cycle

A small and specific subpopulation of cardiomyocytes of the myocardium, termed "pacemaker cells," has the ability to generate action potentials locally in the sinoatrial [30]. Once initiated this depolarization signal is propagated, traveling down the atrium, to the AV node, where it is able to pass into the bundle of His and finally into the subendocardium via a specialized collection of conducting tissue called the "Purkinje fibers." The Purkinje fibers are essential to heart conduction, maintaining a consistent rhythm, as they are able to conduct cardiac action potentials more efficiently and faster than their smaller cardiomyocyte counterparts [31]. This is due to their unique structural organization. In comparison to cardiomyocytes, Purkinje fibers have a larger number of mitochondria and fewer myofibrils. Electrocardiography serves as a medical tool to monitor patterns in heart conduction and can be broken down into four distinct phases: the P wave, representing atrial depolarization; a QRS complex, which is characteristic of ventricular depolarization; the T wave, which shows ventricular repolarization; and finally the U wave, a pattern that denotes papillary muscle (ventricle muscles) repolarization [21, 31]. Diastolic and systolic cycles act in concert for blood perfusion. The left and right atria of the heart are filled by veins in the first stage of diastole, while the left and right ventricles are relaxed. In the second stage, contraction of the right and left atria by the myocardium results in the pumping of blood into the right and left ventricles via the AV valves. In the first stage of systole, both ventricles contract simultaneously, pushing blood, where pulmonary arteries are fed by the right ventricle, while the aorta is fed by the left ventricle. Systolic and diastolic blood pressure are measured from the second stage of the systole and the diastole, or the pressure of the heart at rest. Normal systolic and diastolic pressure ranges are between 80–120 and 60–80, respectively [32].

4.3.2 Blood Mechanics

Blood is composed of three components: red blood cells (or erythrocytes); white blood cells, containing leukocytes and platelets; and plasma, an aqueous solution that contains ionic solution, water, and proteins. The total blood constituents can be fractioned as follows: 45% red blood cells, 0.18% white blood cells, and the remaining plasma. Since 90% of the plasma is made up of water, the non-Newtonian fluid properties of blood can be attributed mainly to erythrocytes, deformable biconcave disks that have an average diameter of approximately 8 μm. Direct measurement of plasma rheological properties has been achieved, although with varying results in that the specific concentrations of albumin, fibrinogen, and globulins do fluctuate. However, general guidelines of plasma protein concentration demonstrate that it consists of ~50% albumin, ~45% globulins, and ~5% fibrinogen [33]. The Newtonian properties of plasma have been measured—with a fluid viscosity of 1.2 centipoise (cP)—making blood a viscous fluid with non-Newtonian behavior.

4.3.3 Cardiovascular Extracellular Matrix Composition

The feedback between physical and chemical cues within the extracellular cardiac niche play an important role in the structure and function of heart and vascular tissue. Bidirectional, inside-out signaling from the cell to the ECM is achieved via transmembrane integrin proteins [34, 35]. The dimerization of integrin α and β subunits physically links intracellular actin-linked proteins such as zyxin, paxillin, talin, vinculin, and α-actinin to ECM residues. The complexity of integrin motifs extends to the ability of multiple or singular ECM ligands to bind to a single integrin receptor. Furthermore, several integrin heterodimers can adhere to a single ECM ligand. Nonetheless, once activated, integrin complexes regulate GTPases such as Rho or Rac, which promotes the assembly of focal adhesion complexes, or kinases such as src kinase and focal adhesion kinase [36].

In general, there are over 20 known members of the integrin family constructed by 8β and 18α subunits. Some of the integrins resident to cardiomyocytes include binding motifs for collagen type I (α3β1), laminin (α1β1, α3β1, and α7β1), and fibronectin (α3β1 and α5β1) [37]. While ECs are anchored to the vessel wall by nearly 20 different ECM proteins, the 2D mesh primarily contains collagen IV, laminin, fibronectin, and proteoglycan perlecan [38]. A thin membrane called the tunica intima, composed of collagen VI and VIII, separates ECs from the tunica media. The tunica media, which is typically thickened in arteries, contains supportive mural cell populations, namely smooth muscle cells, which are found in large arteries and veins, and pericytes, which reside in small capillaries. The specific ECM contribution to vasculature is dependent on tissue function. Vascular tone is mediated, in part, by the elastic nature of the ECM proteins fibrillin and elastin. Where vessel elasticity is not essential, as in the case of vascularized muscle tissues, whose role is to distribute blood to different organs, the ECM is enriched with concentric layers of smooth muscle cell sheaths.

Experiments utilizing atomic force microscopy demonstrated that developing hearts in chicken embryos, specifically 36 to 408 hours after fertilization, show a steady increase in elastic moduli, ranging from 0.9 kPa to 8 kPa, during development [39]. These changes in stiffness of the developing heart coincide with changes in the ECM composition, identified through time-traced studies using quantitative polymerase chain reaction. Here, heart development was preceded by increased collagen expression, with maturation leading to decreasing laminin and fibronectin expression.

The combinatorial roles of elastic fibrous proteins such as fibrin and collagen, as well as adhesive glycoproteins such as laminin and fibronectin, in the basal laminae aid in the proper configuration of cells within the heart. For example, the orientation of cardiomyocytes is dictated by the fibrillar architecture of the surrounding collagen as assembled primarily by cardiac fibroblasts, ensuring regularity in adjacent sarcomere organization, which supports systolic performance [40]. In addition, collagen serves as an electrically conductive material, allowing action potential propagation extending from adjacent cardiomyocytes to the cells of the AV node, bundle branches, and Purkinje fibers.

Increasing research has outlined the role of glycosaminoglycans (GAGs) and matrix proteoglycans in forming the early structures of the heart [1, 41–43]. Gradients of hyaluronic acid (HA) and fibronectin 1 (Fn1) in particular help guide heart field migration, formation of early heart tubes, and chamber septation [1, 42, 44–46]. Gradually, as migration and proliferation occur, heart structures stabilize, and postnatal hemodynamics begins to put biomechanical strain on the myocardium. As a result, cardiac fibroblasts and other supporting cell types exchange the softer, growth-factor-rich gels of HA with stiffer, highly cross-linked, aligned fibers of collagens I and III, supported by collagen IV and laminin in particular. The adult heart is highly fibrillar (particularly in the ventricles), guiding anisotropic conduction, but relatively poor in fibronectin, GAGs, and other proteoglycans, which guide effective wound healing in other tissues, leading to the fatty and scar-like responses to insults such as infarction [47, 48].

Cardiac β-adrenergic responses, influencing action potential morphology, calcium metabolism, force development, and tension, are all mediated by laminin-mediated integrin activation [49]. Integrin signaling is equally important in the functionality of other cell types within the heart, such as cardiac fibroblasts [37, 43, 47, 50]. Fibronectin and osteopontin adhesion are mediated by α5β1 expression by cardiac fibroblasts. Additionally, the expression of α5β1, α5β3, and α5β5 integrins drives cardiac fibroblast adhesion to vitronectin, fibronectin, and osteopontin. Novel in vitro and in vivo experiments have begun to uncover the role of integrin expression and ECM composition in cardiac development. In comparison to adult cardiac fibroblasts, embryonic cardiac fibroblasts are able to stimulate the proliferation of cardiomyocytes in an ECM-dependent manner [51]. Using messenger RNA profiling techniques, Ieda et al. found embryonic cardiac fibroblasts have enhanced expression of proteins, such as fibronectin, and collagens tenascin C and periostin when compared to adult cardiac fibroblasts. In coculture experiments, where cardiomyocytes were cultured in the presence embryonic cardiac fibroblasts, inhibition of cardiac proliferation was achieved when fibroblast expression of Fn1 and/or collagen (Col3a1) was temporarily knocked down through small interfering RNA treatment. Similarly, when cardiac β1 integrin expression was inhibited, proliferation was suppressed [51].

4.4 Engineering Approaches to Studying Mechanotransduction in Cardiovascular Development

Progressive heart failure results from infarction, due to the lack of functional regeneration of the myocardium, leading to aneurysmal thinning and scarring. This can be attributed to the lack of stemness within cardiomyocytes and their inability to proliferate. Tissue engineering approaches to augment the function of damaged cardiac tissue include implementing cellular grafts that either induce an angiogenic response or supplement the mechanical requirement of the infarcted wall. Both of these approaches are key in re-establishing the structure for proper ventricular function. After infarction, engrafted cardiomyocytes of fetal or neonatal origin have been shown to generate new functional tissue separate from the injured scar tissue. In addition to the lack of functional integration in diseased sites, many of the implanted cardiomyocytes died upon implantation, clearly demonstrating the need for structural support for viable engraftment [52].

The limited regenerative capacity of adult heart muscle, attributed variously to insufficient proliferation and lack of stemness in the resident populations of the myocardium, continues to motivate technologies for cardiac regeneration. Bare injection or implantation of preparations of differentiated cardiomyocytes (such as injections, clusters, aggregates, spheroids, or cell sheets) for heart regeneration is hampered due to diminished postimplantation viability and lack of functionality. Cardiomyocytes differentiated from human stem cells currently reach only a fetal-like maturation phenotype, with poor sarcomere organization, immature calcium cycling, pacemaker-like automaticity, and both low expression and poor localization of appropriate gap junctions and mature potassium channels in particular. This leads to the poor functional integration that is often seen between stem cell–derived cardiomyocytes and remaining viable native tissue, with poor electromechanical cell-cell coupling, and also increases concerns that therapy with these preparations might potentially be arrhythmogenic in and of themselves. To this end, engineering approaches to the in vitro conditioning and in vivo delivery of cardiovascular cell populations have been highlighted as among the most promising strategies to improve viability, engraftment, and functionality of cardiomyocytes and other necessary cell types; once surgically implanted, these biomaterials also facilitate host cell infiltration, further ensuring that transplanted cells can functionally integrate into and augment the function of the diseased tissue (Fig. 4.1).

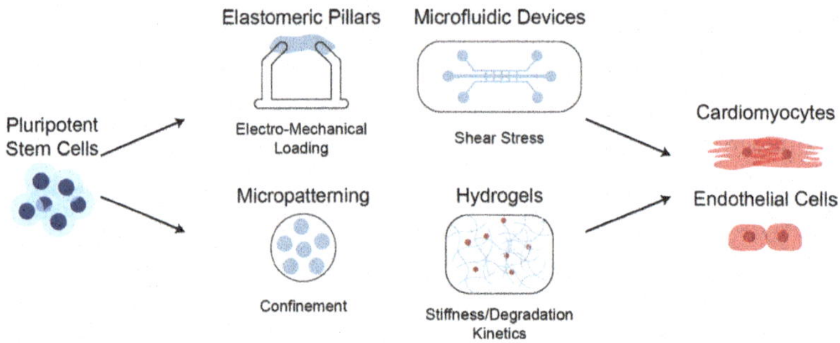

Figure 4.1 Engineering tools to mechanically drive cardiovascular specification from pluripotent stem cells.

4.4.1 Cell Sources

Tissue regeneration is facilitated in part by a complex cross talk between the extracellular milieu of stem, progenitor, and mature cells. This environment, rich in chemical cues and structural support, acts to drive and support states of self-renewal, differentiation, and quiescence. A large facet of tissue engineering approaches focuses on co-delivering cardiovascular cells with regenerative capacity alongside biomaterials that recapitulate this nurturing environment to sites of disease. Once homed in on the site of injury, the goal is for this construct to integrate within the diseased site, where the delivered cells can act to replace or augment the damaged tissue. Due to the limited growth and regenerative potential of somatic cardiovascular tissue, a range of progenitor and pluripotent sources have emerged as viable alternatives.

4.4.1.1 Pluripotent cells

Both the isolation of human embryonic stem cells (hESCs) in 1998 [53] and the advent of human-induced pluripotent stem cells (hiPSCs) technology nearly a decade later, where genetic reprogramming of somatic cells takes on a nascent state [54], have sparked a wave of new opportunities for biomedical research and tissue regeneration studies. These opportunities have become available as pluripotent stem cells (PSCs) have the unique ability to self-renew indefinitely while maintaining full differentiation capacity even toward extraembryonic tissue. As such, differentiation protocol advances toward the various constitutions of the cardiovascular system [55–69], with improved efficiency, specificity, and maturation states, continue to position PSCs as a revolutionary tool that can address major limitations in basic research and regenerative therapies. Methods to reproducibly expand and mass-produce patient-derived stem cells and their derivatives continue to be a goal to address the organ donor shortage crisis and the lifelong immunosuppression currently needed for transplantation patients.

Throughout this chapter, we will emphasize engineering strategies to mimic mechanical cues present in the extracellular milieu, focusing on the forces re-created in the lab that modulate the pluripotent and differentiation capacity of PSCs.

4.4.1.2 Mesenchymal-derived stem cells

Although multipotent mesenchymal stem cells (MSCs) have limited differentiation capacity, they are not restricted from mesodermal lineages and can be specified to a range of cardiovascular subtypes, including ECs [70], smooth muscle cells [71], and various myocyte lineages, such as cardiomyocytes [72]. Similarities between MSC and stromal populations, including pericytes [73] and adventitial cells [74], further support the use of MSCs in vascular regeneration strategies. MSCs do not have a singular identifying factor, which poses as a potential limitation in their widespread use, but in general, MSCs share a class of markers, including CD14, CD45, CD166, CD105, CD29, and STRO-1 [75]. However, MSCs are resident in almost all tissues, sequestered in tissues ranging from hair follicles and tooth roots to adipose, umbilical, and bone marrow tissue. In addition to their wide availability, MSCs are immunogenic, lacking major histone class II expression, and are immunosuppressive when challenged with cytokines from activated leukocytes [76].

4.4.1.3 Progenitor cells

A minute population of functionally heterogenous circulating endothelial progenitor cells (EPCs) participates in vasculogenic events after homing to sites of ischemic injury [77, 78]. Like MSCs, a specific marker cannot be used to distinguish this unique population of progenitor cells. Instead, a panel of surface markers is required for their proper identification. While many blood-derived progenitor cells can contribute to vascular repair, many, such as colony forming unit–Hill cells, take on more of a hematopoietic signature when compared to primary endothelium. However, resident evidence suggests that the blood derivatives erythro-myeloid progenitors have both endothelial and hematopoietic potential, contribute to vascularization, and persist through adulthood and can be found in heart tissue [79]. Human isolation of these cells has yet to be achieved, but stem cell technologies can be used to generate these newly identified cells. However, robust studies have been conducted with endothelial colony forming cells and have been extensively reviewed elsewhere [80]. Importantly these isolates can now be derived from hPSCs [81]; demonstrate endothelial characteristics, including the ability to form lumenized capillary tubes; align to shear stress [82]; and stably integrate and anastomose into implanted tissue.

Unlike EPCs, the identification and characterization of stem cells residing in the heart have and continued to be surrounded by mystery and controversy. Although renewal of cardiomyocytes has been shown (which is less than 1% annually) [83], evidence suggesting higher rates or turnover has been repeatedly refuted and/or retracted. Of intense debate is the notion that cKit-positive cardiac stem cells can give rise to skeletal muscle, endothelial cells, and myocyte cells in vivo. While these claims have been poorly corroborated [84] and are under heavy debate, there is still the need to understand how these populations can be manipulated in vitro for understanding cardiac differentiation capacity [85].

4.4.2 Extracellular Matrix Regulation of Cardiovascular Development and Regeneration

4.4.2.1 Decellularized tissue

The current last-resort alternative in end-stage heart failure is allo-transplantation, requiring life-long administration of immunosuppressive drugs to prevent graft rejection and a relatively limited lifespan of the donor heart (as compared to other organs often transplanted). Top-down tissue engineering approaches for cardiovascular regeneration often involves delivering therapeutic cells in engraftable materials, with the hope that these cells are able to either (i) engraft, repopulate, and function, and/or (ii) develop a transient paracrine milieu to guide host cell infiltration and response. One paradigmatic and promising approach is using decellularized tissue as a scaffold after removing immunogenic donor cells through a variety of techniques, including mechanical abrasion, enzymatic digestion, or administration of ionic and non-ionic detergent solutions. With careful application of decellularization techniques, the physical and chemical cues resident to the native ECM are largely retained, permitting cells to functionally engraft. Relatedly, in peripheral vascular tissue engineering applications, detergents such as sodium dodecyl sulfate have shown promise for complete intra- and extraluminal decellularization. With this technique, these decellularized structures have been used "off-the-shelf" for arterio-venous fistulas, dialysis grafts, and arterial bypass surgery; in the latter case, these grafts maintained patency 2 weeks after engraftment through structural robustness and lack of inflammation [86].

4.4.2.2 Natural extracellular matrices

Matrigel, a xenogeneic conglomeration of ECM proteins from the Engelbreth–Holm–Swarm mouse sarcoma cell line has been extensively used in human pluripotent stem cell culture and differentiation techniques. Although its components and the proportions thereof are largely undefined and slightly variable batch to batch, this laminin-rich mixture of proteins has been shown to enhance cardiac differentiation potential and is widely used as a tissue-culture vessel coating. However, its uses have also expanded to include 3D applications, including organoid formation and vascular models, that take advantage of its gel-like properties. For example, using a Matrigel sandwich technique, where a thin monolayer of high-concentration Matrigel is overlaid on differentiating stem cells, mesoderm differentiation efficiency and downstream cardiac specification were improved.

This change in robust differentiation was found to be mediated by an ECM-driven epithelial-to-mesenchymal transition, where differentiated populations exposed to Matrigel on the apical membrane led to an increase in the ratio between N-cadherin (mesenchymal) and E-cadherin expression (epithelial) [68].

Beyond Matrigel-based matrices, research groups have isolated native cardiac tissue directly for the purpose of creating a biomimetic scaffold. This has been achieved, for example, through the decellularization of porcine hearts, through a series of acidic and enzymatic digestion processes. Once isolated, the samples were lyophilized as pure or hybrid hydrogels, doped with varying concentrations of collagen I. When hESCs were differentiated toward cardiac lineages in these hybrid hydrogels, it was found that the ECM composition had profound effects on cardiac differentiation potential and downstream functionality, even in the absence of supplemental growth factors. Cardiac troponin T marker expression, measured through reverse transcription polymerase chain reaction (RT-PCR), demonstrated that after 8 and 12 days of differentiation, hESCs cultured in hydrogels with high ECM content, namely those composed of 75% decellularized cardiac tissue and 25% collagen I, differentiated with higher efficient when compared to 75% collagen I composites or pure collagen I matrices. Immunofluorescence investigation of connexin-43 (Cx-43) and troponin 1 further demonstrated the role of high native ECM composition in cardiac maturation, where only 75/25 hydrogels led to robust troponin I striations and organized Cx-43 distribution. Interestingly, this was only evident in conditions of absence of traditionally used maturation growth factors VEGF and DKK1 [87].

These matrices can also be generated in vitro by cells themselves. Leveraging the notion that embryonic cardiac fibroblasts stimulate cardiomyocyte proliferation. Baharvand et al. demonstrated that this ECM also influences the maturation of stem cell–derived cardiomyocytes. Through ethylenediaminetetraacetic acid (EDTA) treatment, embryonic fibroblasts can be selectively removed in vitro, leaving the ECM they secrete intact. This ECM, termed "cardiogel," was able to sustain cardiomyocyte functionality, relative to Matrigel, and led to earlier chronotropic maturation, evidenced by ultrastructure analysis [88].

Additional studies have uncovered the role of ECM in the propagation of differentiated cues toward cardiac specification. For example, the role of ascorbic acid in enhancing cardiac specification from embryonic stem cells was found to be attributed to increased collagen synthesis [89]. Downstream maturation of stem cell–derived cardiomyocytes is also driven by ECM presentation.

4.4.2.3 Synthetic matrices

In addition to naturally derived biomaterials, designer hydrogel matrices, which capture the physical cues resident to native ECM, have proven to be powerful tools for studying mechanotransduction in cardiovascular development and regeneration. As opposed to natural ECM, synthetic hydrogels are not limited by batch-to-batch variability and can be designed to have specific integrin-binding motifs, degradation kinetics, topography, porosity, and matrix stiffness. In addition, these synthetic systems permit ECM remodeling by resident cells, mimicking the native ECM. For example, hydrogels can be engineered to mimic physiologically relevant stiffness through cross-linking chemistry. Specifically, one can tailor the mechanical properties of a hydrogel by altering the concentration of polymer and cross-linking agents. Li et al. demonstrated the benefits of engineering a hydrogel scaffold with $\alpha 3/\alpha 5$ $\beta 1$ integrin-binding sites, which promotes a dense network of vasculature with extensive sprouting and branching and the ability to achieve anastomosis [90]. By achieving such a platform, they were able to reduce the amount of VEGF needed for delivery within the scaffolds in vivo, allowing the development of nontortuous and nonleaky blood vessels (a necessity in cerebrovascular applications in particular).

The creation of hydrogels that are photopolymerizable has also been a major advancement for achieving mechanical stability during vascularization—for example, a gelatin methacrylate hydrogel demonstrated superior mechanical stability, the support of dense lumen structures, and an ability to be modulated to control cell behavior [91]. HA hydrogels have demonstrated particularly beneficial properties. In work by Hanjaya-Putra et al., tunable HAs demonstrated the ability to support endothelial colony forming cells in the creation of vascular networks (vacuole and lumen formation due to the presence of integrin-binding sites $\alpha 5\beta 1$ and $\alpha 5\beta 3$ and further branching enabled by hyaluronan degradation) [92]. This work has been extended to support the creation of self-organized, robust vascular networks by pluripotent stem cell–derived ECs and pericytes [93].

These synthetic systems have demonstrated applications beyond vascular development and in the manipulation of cardiomyocytes and cardiac tissue. To mimic the native stiffening environment of developing cardiac tissue, Young et al. demonstrated the role of dynamic hydrogel modulation on cardiomyocyte differentiation in vitro. Using poly(ethylene glycol) diacrylate (PEG-DA) as a cross-linker and thiolated-HA, the material was engineered to stiffen from ~2 kPa to ~8 kPa over a 456-hour period. Embryonic cardiomyocytes isolated and plated on these hydrogel surfaces showed increased cardiac maturity through myofibril alignment and decreased expression of precardiac marker NKX-2.5 [39]. Using a similar platform, PEG-DA fibrinogen hydrogel composites were used to access the effects of substrate moduli and spontaneous contraction of cardiomyocytes. On increasing the concentration of fibrinogen, the shear moduli of the construct increased. The result of increased stiffness was a decrease in the percentage and synchrony of neonatal cardiomyocytes contraction [94].

In regard to embryonic development toward cardiovascular lineages, substrates ranging in compliance have been instrumental tools in understanding the role of mechanical stimulation on early fate decisions. Polyacrylamide hydrogels and polydimethylsiloxane (PDMS) have been exploited in a wide range of systems to study the role of physiologically relevant stiffnesses. Several examples have demonstrated that mesoderm differentiation efficiency is enhanced when PSCs are subjected to soft surfaces. Mechanistic evaluation has determined that soft substrates mediate Wnt signaling in differentiating hPSCs, through B-catenin degradation [95], and this is related to the mechanosensitive transcriptional regulator Yes-associated protein/transcriptional co-activator with PDZ-binding motif (YAP/TAZ) [96].

Specifically, hPSCs undergo rapid differentiation and endothelial maturation when the mesoderm specification is primed on compliant surfaces when compared to supraphysiologically stiff substrates [96].

4.4.2.4 Oxygen tension and mechanotransduction

Oxygen tension is a potent driver of vascular morphogenesis, a phenomenon during de novo vasculogenesis or vascular remodeling in adulthood via angiogenesis. A variety of engineering tools have been developed to mimic dissolved oxygen levels of tissue (1%–5%) and blood vessels (5%–7%). In addition to a number of angiogenic genes that are upregulated during hypoxia, including the hypoxia-inducible factors (HIFs), VEGF, and angiopoietin-2, many elements involving ECM remodeling are altered.

Moderate hypoxic exposure (~5%) coaxes induced pluripotent stem cells (iPSCs) to undergo enhanced endothelial differential potential. Hypoxic exposure after mesodermal differentiation is inefficient in stimulating endothelial lineage specification, suggesting the temporal nature of vascular differentiation [64]. In addition to the role of hypoxic in vascular differentiation, low oxygen tension coaxes an angiogenic response by stimulating both mature ECs and EPCs to increase deposition of ECM proteins such as laminin, collagen IV, collagen I, and Fn1 [97]. This hypoxia-induced ECM production is crucial for development and wound healing. Interestingly, ECM deposition was accompanied by a changing cytoskeletal structure in mature ECs and EPCs. For example, disruption of F-actin assembly diminishes the ability for ECs under hypoxic stimulation to produce fibronectin.

To better understand the role of hypoxic stimulation in vascular development and regeneration, novel biomaterials, with tunable properties, that mimic the three dimensionality of native tissue have been developed. Using gelatin as a polymer backbone, functionalized with ferulic acid via a carbodiimide-mediated reaction, permits the formation of a hypoxic gel via a laccase-driven cross-linking reaction that consumes oxygen [98]. In this system, EPCs were encapsulated within the hydrogel and were found to undergo robust vasculogenesis compared to normoxic gelatin hydrogels. Expectedly HIF isoforms were upregulated in hypoxic culture, but ECM remodeling factors, which are important mediators of morphogenesis, were simultaneously stimulated. Specifically, cells in hypoxic hydrogels had an increased expression of matrix metalloproteinase (MMP)-1, MMP-2, and MT1-MMP compared to normoxic controls. Acellular implantation of these oxygen-controlled materials stimulated host neovascularization, demonstrating the importance of oxygen tension in vascular regeneration. This laccase-mediated chemistry has served as a powerful tool for understanding the role of oxygen in a variety of natural and synthetic polymer backbones and will serve as an important platform for in vitro control of the microenvironment to study vascular regeneration [99–101].

4.4.3 BioMEMS

The manipulation of fluids on the scale of tens to hundreds of microns has been designed to mimic and generate small capillary vascular beds for transplantation and high-throughput drug discovery. Leveraging technology from the semiconductor manufacturing industry, bioMEMS have served as a powerful tool to create micron-scale platforms that allow the unprecedented control of mechanical forces that are present during cardiovascular development. To fabricate these devices, photomasks with physiologically mimicking geometric patterns can be transferred onto a silicon master mold via photolithography with light-sensitive photopolymers such as SU-8 epoxy, that cross-links on silicon substrates in the presence of ultraviolet (UV) light.

A variety of steps control the axial heights of the micropatterned features, including the duration of SU-8 resin development or the height at which the polymer is spun onto the silicon substrate. Once a patterned substrate is made, the features can be transferred via replica molding, where an elastomeric material such as PDMS conforms to the shape and can be removed at ease. The resulting relief features can be used for the fabrication of many devices, including micropillars, micropatterns, and microfluidics, whose utilization for vascular tissue engineering applications has been extensively reviewed [102].

4.4.3.1 Microfluidic platforms

Microfluidic in vitro models have permitted the mechanistic evaluation of laminar shear regimes [27] on normal iPSC-derived endothelial homeostasis as well the role of turbulent and oscillatory stress in endothelial inflammation [103, 104]. Arterial specification from stem cell–derived ECs [105] has been achieved with the application of laminar shear stress and has served as an important tool to study endothelial maturation capacity and function. With the use of several differentiation protocols and somatic sources, the functionality of iPSC-derived ECs was appraised using a peristaltic pump integrated with a 2D microfluidic system. iPSC-derived ECs with the capacity to form primary cilia, a mechanosensitive microtubule–based organelle, correlated with nascent EC alignment after shear application and the ability to uptake exogeneous calcium

[82, 106]. On the other hand, powerful 3D microfluidic platforms recapitulate the native EC-matrix interactions and have been used to investigate Notch-mediated endothelial permeability [107] as well as the roles of matrix composition, interstitial flow, and chemical cues during angiogenesis [108–111].

4.4.3.2 Micropatterned tools

Physiochemical instructions orchestrate coordinated expansion and self-assembly into 3D architectures from a cluster of pluripotent cells, giving rise to three distinct germ layers that eventually form an entire organism. While the kinetics of this process varies tremendously between across species, interactions between cells and their surrounding ECM, as well as cavitation, polarization, and migration are conserved. Highly innovative and quantitative metrics have evolved to capture the basics of these morphogenic events in vitro and have created an entire new interdisciplinary developmental biology field [112]. What is unique to the observation of self-organizing human pluripotent tissue, to structures that mimic natural development, is the necessity for both potent morphogenic growth factor gradients and physical cues. Historically, capturing developmental self-organization from PSCs, through embryoid bodies (EBs), resulted in a chaotic assembly of cells representing all three germ layers; however, those events relied on encapsulation of ill-defined ECM compositions to induce polarization events. On the contrary, controlled presentation of ECM through micropatterned technologies in two dimensions has led to the generation of well-organized gastruloids [113]. Coupled with immunofluorescence staining and computational tools, high-throughput modalities have been created to monitor the driving forces behind these organization events. In particular, micropatterned islands have been used to investigate differentiation kinetics [114], and the contribution of E-cadherin-mediated cell-cell interactions toward the loss of pluripotent marker expression, in hiPSCs differentiating toward mesodermal lineages [115].

Under defined chemical conditions, micropatterned domains have been leveraged to recreate, elucidate, and computationally model Alan Turing–like BMP/WNT/Nodal morphogen gradients in human embryonic development [113, 116–118], leading to well-defined gastrulation-like spatial organization.

Beyond a utility to study early organization events, micropatterned geometries have been used to understand the role of confinement in early lineage segmentation (how those cues restrict downstream specification of cardiovascular lineages). ECM availability biases hiPSC-derived bipotent progenitors toward an endothelial fate over pericyte specification [119], and disruption of key mediators of cytoskeletal tension disrupts early mesoderm organization, with consequences in the downstream self-assembly of cardiac [120] and vascular tissues [121]. These systems are not restricted to 2D domains. One example is the use of PEG microwells to control the size of EBs by microfabricating PDMS models (150–450 μm in diameter). Here, EBs transferred from 150–300 μm microwells onto Matrigel-coated substrates substantially formed more angiogenic sprouts when compared to EBs grown in 450 μm microwells. Studies like these demonstrate how the control of homotypic interactions between cells can modulate maturation tendencies in cardiovascular development [112]. These technologies permit high-through toxicity studies and are amenable to new image analysis and computation tools that will accelerate basic research studies.

4.4.4 3D Printing Technology

The first example of 3D printing, an extension of 2D printing technologies where successive layers or materials could be organized into more complex structures, was pioneered in the mid-1980s and was termed "stereolithography." Charles W. Hull, who first described this process, was able to build 3D architecture by photosensitive materials that are cross-linkable by UV light [122]. Building upon this layer-by-layer approach where liquid resin is photopolymerized, prior to the addition of a new layer, two-photon polymerization, digital projection lithography, and continuous liquid interface production have emerged as faster techniques with higher spatial resolutions. Today an array of modalities are used to mimic the complex structure of native tissue, where cells, signaling cues, and ECM components are spatially organized with printing technology.

Thermally extruded carbohydrate glass, built as intertwined lattices, was embedded in a range of prepolymerized ECM composites of both natural and synthetic origin containing viable cells. After polymerization through various means, dependent on the casted biomaterial, the lattices could be readily dissolved by cell media, leaving an open vascular network predefined by the printing architecture. Intriguingly, these lattices could be seeded with ECs, and perfused, along cells that were embedded in the bulk matrices [123]. Porosity of alginate 3D-printed gels was shown to be a key design criterion in sustaining the viability of human fetal cardiomyocyte progenitor cells.

Functionalization of these 3D-printed constructs with the RGD-binding motif permitted cell elongation and spreading, leading to a significant increase in mature cardiac marker expression [124].

Iterations of the seminal work done by Feinberg et al. [125], where anisotropic flexible PDMS films were exploited to create muscular thin film actuators, are now used for drug screening agents that induce cardiotoxicity or have the potential for therapeutic application [126, 127]. Through multimaterial 3D printing technology, anisotropic cardiac tissue can be seeded on a flexible film with embedded strain sensors that can be read out electrically [128]. In a similar manner, a sequential-spin-coating methodology was used to create a facile version of the 3D-printed cantilever array [129]. This platform also permitted the addition of an endothelial barrier, creating a drug screen platform that was able to recapitulate drug transport in normal and inflamed vascular tissue.

4.5 Conclusions and Future Directions

Globally, cardiovascular diseases remain the leading cause of death. Unfortunately, the current standard intervention involves surgical replacement or use of palliative drugs for symptom alleviation. These treatment routes do not address the underlying mechanism behind these ailments involving direct injury to the myocardium, supporting ECM or vasculature. Cell-based therapy approaches seek to introduce regenerative cell populations that can act to augment function by inducing self-repair mechanisms or replacing dead or damaged cells. A range of adult, progenitor, and stem cell sources have been tested, with varying levels of success. Self-renewing pluripotent sources with cardiac and vascular maturation potential can be produced at an industrial scale (~750 million) and have been demonstrated to remuscularize infarcted regions in macaque studies [130]. While it is increasingly evident that cell sourcing is critical to the clinical success of cell therapies [131], their successful delivery into damaged tissue is paramount. Hydrogels pose as a delivery vehicle to increase the survival and engraftment of stem cell–derived cardiovascular cells. Novel engineered injectable natural and synthetic scaffolds, with carefully designed physical and chemical cues, can support the integration and sustained function of stem cell cardiovascular derivatives [132]. These techniques, coupled with an increasing understanding of the necessity of mechanical cues in the maturation of stem cell–derived cardiovascular cells [133], will bolster the promise of cell therapy.

References

1. Hanson, K. P., et al. (2013). Spatial and temporal analysis of extracellular matrix proteins in the developing murine heart: a blueprint for regeneration. *Tissue Eng Part A*, **19**, pp. 1132–1143.
2. Van Vliet, P., Wu, S. M., Zaffran, S. and Puceat, M. (2012). Early cardiac development: a view from stem cells to embryos. *Cardiovasc Res*, **96**, pp. 352–362.
3. Davies, S. W. and Wedzicha, J. A. (1993). Hypoxia and the heart. *Br Heart J*, **69**, pp. 3–5.
4. Harris, I. S. and Black, B. L. (2010). Development of the endocardium. *Pediatr Cardiol*, **31**, pp. 391–399.
5. Brutsaert, D. L. (2003). Cardiac endothelial-myocardial signaling: its role in cardiac growth, contractile performance, and rhythmicity. *Physiol Rev*, **83**, pp. 59–115.
6. Rochais, F., Mesbah, K. and Kelly, R. G. (2009). Signaling pathways controlling second heart field development. *Circ Res*, **104**, pp. 933–942.
7. Takao, A., Ando, M., Cho, K., Kinouchi, A. and Murakami, Y. (1980). *Etiology and Morphogenesis of Congenital Heart Disease*. pp. 253–269.

8. Foglia, M. J. and Poss, K. D.(2016). Building and re-building the heart by cardiomyocyte proliferation. *Development*, **143**, pp. 729–740.

9. Kimura, W., et al. (2015). Hypoxia fate mapping identifies cycling cardiomyocytes in the adult heart. *Nature*, **523**, pp. 226–230.

10. Nakada, Y., et al. (2017). Hypoxia induces heart regeneration in adult mice. *Nature*, **541**, pp. 222–227.

11. Puente, B. N., et al. (2014). The oxygen-rich postnatal environment induces cardiomyocyte cell-cycle arrest through DNA damage response. *Cell*, **157**, pp. 565–579.

12. Tseliou, E., et al. (2014). Angiogenesis, cardiomyocyte proliferation and anti-fibrotic effects underlie structural preservation post-infarction by intramyocardially-injected cardiospheres. *PLoS One*, **9**, pp. e88590.

13. Sassoli, C., et al. (2011). Mesenchymal stromal cells affect cardiomyocyte growth through juxtacrine Notch-1/Jagged-1 signaling and paracrine mechanisms: clues for cardiac regeneration. *J Mol Cell Cardiol*, **51**, pp. 399–408.

14. Sedmera, D. and Thompson, R. P. (2011). Myocyte proliferation in the developing heart. *Dev Dyn*, **240**, pp. 1322–1334.

15. Xin, M., Olson, E. N. and Bassel-Duby, R. (2013). Mending broken hearts: cardiac development as a basis for adult heart regeneration and repair. *Nat Rev Mol Cell Biol*, **14**, pp. 529–541.

16. Yannarelli, G., et al. (2014). Donor mesenchymal stromal cells (MSCs) undergo variable cardiac reprogramming in vivo and predominantly co-express cardiac and stromal determinants after experimental acute myocardial infarction. *Stem Cell Rev Rep*, **10**, pp. 304–315.

17. Chan, S. K., et al. (2016). Development of bipotent cardiac/skeletal myogenic progenitors from MESP1+ mesoderm. *Stem Cell Rcp*, **6**, pp. 26–34.

18. Rahbarghazi, R., et al. (2013). Juxtacrine and paracrine interactions of rat marrow-derived mesenchymal stem cells, muscle-derived satellite cells, and neonatal cardiomyocytes with endothelial cells in angiogenesis dynamics. *Stem Cells Dev*, **22**, pp. 855–865.

19. Condorelli, G., et al. (2001). Cardiomyocytes induce endothelial cells to trans-differentiate into cardiac muscle: implications for myocardium regeneration. *Proc Natl Acad Sci U S A*, **98**, pp. 10733–10738.

20. Gokcinar-Yagci, B., Uckan-Cetinkaya, D. and Celebi-Saltik, B. (2015). Pericytes: properties, functions and applications in tissue engineering. *Stem Cell Rev*, **11**, pp. 549–559.

21. Heynen-Genel, S. and Mercola, M. (2014). Cardiac, smooth, and skeletal muscle electrophysiology II. *Biophys J*, **106**, pp. 720a.

22. van Blitterswijk, C., et al. (2008). *Tissue Engineering* (Academic Press), pp. 760.

23. Flamme, I., Frolich, T. and Risau, W. (1997). Molecular mechanisms of vasculogenesis and embryonic angiogenesis. *J Cell Physiol*, **173**, pp. 206–210.

24. Patel-Hett, S. and D'Amore, P. A. (2011). Signal transduction in vasculogenesis and developmental angiogenesis. *Int J Dev Biol*, **55**, pp. 353–363.

25. le Noble, F., et al. (2004). Flow regulates arterial-venous differentiation in the chick embryo yolk sac. *Development*, **131**, pp. 361.

26. Lalor, P. F., Lai, W. K., Curbishley, S. M., Shetty, S. and Adams, D. H. (2006). Human hepatic sinusoidal endothelial cells can be distinguished by expression of phenotypic markers related to their specialised functions in vivo. *World J Gastroenterol*, **12**, pp. 5429–5439.

27. Satchell, S. C. and Braet, F. (2009). Glomerular endothelial cell fenestrations: an integral component of the glomerular filtration barrier. *Am J Physiol Heart Circ Physiol*, **296**, pp. F947–F956.

28. Wanjare, M., Kusuma, S. and Gerecht, S. (2013). Perivascular cells in blood vessel regeneration. *Biotechnol J*, **8**, pp. 434–447.

29. Iivanainen, E., et al. (2003). Angiopoietin-regulated recruitment of vascular smooth muscle cells by endothelial-derived heparin binding EGF-like growth factor. *FASEB J*, **17**, pp. 1609–1621.

30. Boyett, M. R., Honjo, H. and Kodama, I. (2000). The sinoatrial node, a heterogeneous pacemaker structure. *Cardiovasc Res*, **47**, pp. 658–687.

31. Ideker, R. E., Kong, W. and Pogwizd, S. (2009). Purkinje fibers and arrhythmias. *PACE*, **32**, pp. 283–285.

32. Liaw, N. Y. and Zimmermann, W.-H. H. (2016). Mechanical stimulation in the engineering of heart muscle. *Adv Drug Deliv Rev*, **96**, pp. 156–160.

33. Woodcock, J. P. (1976). Physical properties of blood and their influence on blood-flow measurement. *Rep Prog Phys*, **39**, pp. 65.

34. Hynes, R. O. (2002). Integrins: bidirectional, allosteric signaling machines. *Cell*, **110**, pp. 673–687.

35. Qin, J., Vinogradova, O. and Plow, E. F. (2004). Integrin bidirectional signaling: a molecular view. *PLoS Biol*, **2**, pp. e169.

36. Price, L. S., Leng, J., Schwartz, M. A. and Bokoch, G. M. (1998). Activation of Rac and Cdc42 by integrins mediates cell spreading. *Mol Biol Cell*, **9**, pp. 1863–1871.

37. Israeli-Rosenberg, S., Manso, A. M., Okada, H. and Ross, R. S. (2014). Integrins and integrin-associated proteins in the cardiac myocyte. *Circ Res*, **114**, pp. 572–586.

38. Rhodes, J. M. and Simons, M. (2007). The extracellular matrix and blood vessel formation: not just a scaffold. *J Cell Mol Med*, **11**, pp. 176–205.

39. Young, J. L. Engler, A. J. (2011). Hydrogels with time-dependent material properties enhance cardiomyocyte differentiation in vitro. *Biomaterials*, **32**, pp. 1002–1009.

40. Baicu, C. F., et al. (2003). Changes in extracellular collagen matrix alter myocardial systolic performance. *Am J Physiol Heart Circ Physiol*, **284**, pp. H122.

41. Abonnenc, M., et al. (2013). Extracellular matrix secretion by cardiac fibroblasts: role of microRNA-29b and microRNA-30c. *Circ Res*, **113**, pp. 1138–1147.

42. Akhyari, P., Kamiya, H., Haverich, A., Karck, M. and Lichtenberg, A. (2008). Myocardial tissue engineering: the extracellular matrix. *Eur J Cardiothorac Surg*, **34**, pp. 229–241.

43. Ross, R. S. and Borg, T. K. (2001). Integrins and the myocardium. *Circ Res*, **88**, pp. 1112–1119.

44. Majkut, S., et al. (2013). Heart-specific stiffening in early embryos parallels matrix and myosin expression to optimize beating. *Curr Biol*, **23**, pp. 2434–2439.

45. Monaghan, M. G., et al. (2016). Endocardial-to-mesenchymal transformation and mesenchymal cell colonization at the onset of human cardiac valve development. *Development*, **143**, pp. 473–482.

46. Nahirney, P. C., Mikawa, T. and Fischman, D. A. (2003). Evidence for an extracellular matrix bridge guiding proepicardial cell migration to the myocardium of chick embryos. *Dev Dyn*, **227**, pp. 511–523.

47. Pentassuglia, L. and Sawyer, D. B. (2013). ErbB/integrin signaling interactions in regulation of myocardial cell-cell and cell-matrix interactions. *Biochim Biophys Acta*, **1833**, pp. 909–916.

48. Piatkowski, T., Muhlfeld, C., Borchardt, T. and Braun, T. (2013). Reconstitution of the myocardium in regenerating newt hearts is preceded by transient deposition of extracellular matrix components. *Stem Cells Dev*, **22**, pp. 1921–1931.

49. Wang, Y. G., Samarel, A. M. and Lipsius, S. L. (2000). Laminin acts via β(1) integrin signalling to alter cholinergic regulation of L-type Ca(2+) current in cat atrial myocytes. *J Physiol*, **526**, pp. 57–68.
50. Stewart, J. A., Gardner, J. D., Brower, G. L. and Janicki, J. S. (2014). Temporal changes in integrin-mediated cardiomyocyte adhesion secondary to chronic cardiac volume overload in rats. *Am J Physiol Heart Circ Physiol*, **306**, pp. H101–H108.
51. Ieda, M., et al. (2009). Cardiac fibroblasts regulate myocardial proliferation through β1 integrin signaling. *Dev Cell*, **16**, pp. 233–244.
52. Murry, C. E., Whitney, M. L. and Reinecke, H. (2002). Muscle cell grafting for the treatment and prevention of heart failure. *J Cardiac Fail*, **8**, pp. S532–S541.
53. Thomson, J. A., et al. (1998). Embryonic stem cell lines derived from human blastocysts. *Science*, **282**, pp. 1145–1147.
54. Takahashi, K., et al. (2007). Induction of pluripotent stem cells from adult human fibroblasts by defined factors. *Cell*, **131**, pp. 861–872.
55. Kusuma, S., et al. (2013). Self-organized vascular networks from human pluripotent stem cells in a synthetic matrix. *Proc Natl Acad Sci U S A*, **110**, pp. 12601–12606.
56. Lian, X., et al. (2012). Directed cardiomyocyte differentiation from human pluripotent stem cells by modulating Wnt/β-catenin signaling under fully defined conditions. *Nat Protoc*, **8**, pp. 162–175.
57. Lian, X., et al. (2014). Efficient differentiation of human pluripotent stem cells to endothelial progenitors via small-molecule activation of WNT signaling. *Stem Cell Rep*, 1–13.
58. Bao, X., et al. (2015). Chemically-defined albumin-free differentiation of human pluripotent stem cells to endothelial progenitor cells. *Stem Cell Res*, **15**, pp. 122–129.
59. Patsch, C., et al. (2015). Generation of vascular endothelial and smooth muscle cells from human pluripotent stem cells. *Nat Cell Biol*, **17**, pp. 994–1003.
60. Mummery, C. L., et al. (2012). Differentiation of human embryonic stem cells and induced pluripotent stem cells to cardiomyocytes: a methods overview. *Circ Res*, **111**, pp. 344–358.
61. Orlova, V. V., et al. (2014). Functionality of endothelial cells and pericytes from human pluripotent stem cells demonstrated in cultured vascular plexus and zebrafish xenografts. *Arterioscler Thromb Vasc Biol*, **34**, pp. 177–186.

62. Orlova, V. V., et al. (2014). Generation, expansion and functional analysis of endothelial cells and pericytes derived from human pluripotent stem cells. *Nat Protoc*, **9**, pp. 1514–1531.

63. Wanjare, M., Kusuma, S. and Gerecht, S. (2014). Defining differences among perivascular cells derived from human pluripotent stem cells. *Stem Cell Rep*, **2**, pp. 561–575.

64. Kusuma, S., Peijnenburg, E., Patel, P. and Gerecht, S. (2014). Low oxygen tension enhances endothelial fate of human pluripotent stem cells. *Arterioscler Thromb Vasc Biol*, **34**, pp. 913–920.

65. Kusuma, S., Facklam, A. and Gerecht, S. (2015). Characterizing human pluripotent-stem-cell-derived vascular cells for tissue engineering applications. *Stem Cells Dev*, **24**, pp. 451–458.

66. Wu, Y.-T., et al. (2015). Defining minimum essential factors to derive highly pure human endothelial cells from iPS/ES cells in an animal substance-free system. *Sci Rep*, **5**, pp. 9718.

67. Kattman, S. J., et al. (2011). Stage-specific optimization of activin/nodal and BMP signaling promotes cardiac differentiation of mouse and human pluripotent stem cell lines. *Cell Stem Cell* **8**, pp. 228–240.

68. Zhang, J., et al. (2012). Extracellular matrix promotes highly efficient cardiac differentiation of human pluripotent stem cells: the matrix sandwich method. *Circ Res*, **111**, pp. 1125–1136.

69. Lian, X., et al. (2012). Robust cardiomyocyte differentiation from human pluripotent stem cells via temporal modulation of canonical Wnt signaling. *Proc Natl Acad Sci U S A*, **109**, pp. E1848.

70. Oswald, J., et al. (2004). Mesenchymal stem cells can be differentiated into endothelial cells in vitro. *Stem Cells*, **22**, pp. 377–384.

71. Tamama, K., Sen, C. K. and Wells, A. (2008). Differentiation of bone marrow mesenchymal stem cells into the smooth muscle lineage by blocking ERK/MAPK signaling pathway. *Stem Cells Dev*, **17**, pp. 897–908.

72. Makino, S., et al. (1999). Cardiomyocytes can be generated from marrow stromal cells in vitro. *J Clin Invest*, **103**, pp. 697–705.

73. Crisan, M., et al. (2008). A Perivascular origin for mesenchymal stem cells in multiple human organs. *Stem Cell*, **3**, pp. 301–313.

74. Crisan, M., Corselli, M. Chen, W. C. W. and Peault, B. (2012). Perivascular cells for regenerative medicine. *J Cell Mol Med*, **16**, pp. 2851–2860.

75. Maleki, M., Ghanbarvand, F., Reza Behvarz, M., Ejtemaei, M. and Ghadirkhomi, E. (2014). Comparison of mesenchymal stem cell

markers in multiple human adult stem cells. *Int J Stem Cells*, **7**, pp. 118–126.

76. Shi, Y., et al. (2010). Mesenchymal stem cells: a new strategy for immunosuppression and tissue repair. *Cell Res*, **20**, pp. 510.
77. Ingram, D. A., et al. (2004). *Identification of a Novel Hierarchy of Endothelial Progenitor Cells Using Human Peripheral and Umbilical Cord Blood*. Vol. 104, pp. 2752–2760.
78. Hirschi, K. K., Ingram, D. A. and Yoder, M. C. (2008). Assessing identity, phenotype, and fate of endothelial progenitor cells. *Arteriosc Thromb Vasc Biol*, **28**, pp. 1584–1595.
79. Plein, A., Fantin, A., Denti, L., Pollard, J. W. and Ruhrberg, C. (2018). Erythro-myeloid progenitors contribute endothelial cells to blood vessels. *Nature*, **562**, pp. 223–228.
80. Yoder, M. C. (2012). Human endothelial progenitor cells. *Cold Spring Harbor Perspect Med*, **2**, pp. a006692.
81. Prasain, N., et al. (2014). Differentiation of human pluripotent stem cells to cells similar to cord-blood endothelial colony–forming cells. *Nat Biotechnol*, **32**, pp. 1151–1157.
82. Smith, Q., et al. (2018). Differential HDAC6 activity modulates ciliogenesis and subsequent mechanosensing of endothelial cells derived from pluripotent stem cells. *Cell Rep*, **24**, pp. 895–908.e896.
83. Bergmann, O., et al. (2009). Evidence for cardiomyocyte renewal in humans. *Science*, **324**, pp. 98–102.
84. van Berlo, J. H., et al. (2014). c-kit$^+$ cells minimally contribute cardiomyocytes to the heart. *Nature*, **509**, pp. 337–341.
85. Hatzistergos, K. E., et al. (2015). cKit$^+$ cardiac progenitors of neural crest origin. *Proc Natl Acad Sci U S A*, **112**, pp. 13051–13056.
86. Schaner, P. J., et al. (2004). Decellularized veinas a potential scaffold for vascular tissue engineering. *J Vasc Surg*, **40**, pp. 146–153.
87. Duan, Y., et al. (2011). Hybrid gel composed of native heart matrix and collagen induces cardiac differentiation of human embryonic stem cells without supplemental growth factors. *J Cardiovasc Transl Res*, **4**, pp. 605.
88. Baharvand, H., Azarnia, M., Parivar, K. and Ashtiani, S. K. (2005). The effect of extracellular matrix on embryonic stem cell-derived cardiomyocytes. *J Mol Cell Cardiol*, **38**, pp. 495–503.
89. Sato, H., et al. (2006). Collagen synthesis is required for ascorbic acid-enhanced differentiation of mouse embryonic stem cells into cardiomyocytes. *Biochem Biophys Res Commun*, **342**, pp. 107–112.

90. Li, S., et al. (2017). Hydrogels with precisely controlled integrin activation dictate vascular patterning and permeability. *Nat Mater*, **16**, pp. 953.

91. Chen, Y.-C., et al. (2012). Functional human vascular network generated in photocrosslinkable gelatin methacrylate hydrogels. *Adv Funct Mater*, **22**, pp. 2027–2039.

92. Hanjaya-Putra, D., et al. (2011). Controlled activation of morphogenesis to generate a functional human microvasculature in a synthetic matrix. *Blood*, **118**, pp. 804–815.

93. Kusuma, S., et al. (2007). Hyaluronic acid hydrogel for controlled self-renewal and differentiation of human embryonic stem cells. *Proc Natl Acad Sci U S A*, **104**, pp. 11298–11303.

94. Shapira-Schweitzer, K. and Seliktar, D. (2007). Matrix stiffness affects spontaneous contraction of cardiomyocytes cultured within a PEGylated fibrinogen biomaterial. *Acta Biomater*, **3**, pp. 33–41.

95. Przybyla, L., Lakins, J. N. and Weaver, V. M. (2016). Tissue mechanics orchestrate Wnt-dependent human embryonic stem cell differentiation. *Stem Cell*, pp. 1–27.

96. Smith, Q., et al. (2017). Compliant substratum guides endothelial commitment from human pluripotent stem cells. *Sci Adv*, **3**, pp. e1602883.

97. Kusuma, S., Zhao, S. and Gerecht, S. (2012). The extracellular matrix is a novel attribute of endothelial progenitors and of hypoxic mature endothelial cells. *FASEB J*, **26**, pp. 4925–4936.

98. Park, K. M. and Gerecht, S. (2014). Hypoxia-inducible hydrogels. *Nat Commun*, **5**.

99. Park, K. M., Blatchley, M. R. and Gerecht, S. (2014). The design of dextran-based hypoxia-inducible hydrogels via in situ oxygen-consuming reaction. *Macromol Rapid Commun*, **35**, pp. 1968–1975.

100. Lewis, D. M., Blatchley, M. R., Park, K. M. and Gerecht, S. (2017). O2-controllable hydrogels for studying cellular responses to hypoxic gradients in three dimensions in vitro and in vivo. *Nat Protoc*, **12**, pp. 1620.

101. Blatchley, M., Park, K. M. and Gerecht, S. (2015). Designer hydrogels for precision control of oxygen tension and mechanical properties. *J Mater Chem B*, **3**, pp. 7939–7949.

102. Smith, Q. and Gerecht, S. (2014). In *Current Opinion in Chemical Engineering* (Elsevier), Vol. 3, pp. 42–50.

103. Shao, J., et al. (2009). Integrated microfluidic chip for endothelial cells culture and analysis exposed to a pulsatile and oscillatory shear stress. *Lab Chip*, **9**, pp. 3118.

104. Khan, O. F. and Sefton, M. V. (2010). Endothelial cell behaviour within a microfluidic mimic of the flow channels of a modular tissue engineered construct. *Biomed Microdevices*, **13**, pp. 69–87

105. Sivarapatna, A., et al. (2015). Arterial specification of endothelial cells derived from human induced pluripotent stem cells in a biomimetic flow bioreactor. *Biomaterials*, **53**, pp. 621–633.

106. Adams, W. J., et al. (2013). Functional vascular endothelium derived from human induced pluripotent stem cells. *Stem Cell Rep*, **1**, pp. 105–113.

107. Polacheck, W. J., et al. (2017). A non-canonical Notch complex regulates adherens junctions and vascular barrier function. *Nature*, **552**, pp. 258.

108. Trappmann, B., et al. (2017). Matrix degradability controls multicellularity of 3D cell migration. *Nat Commun*, **8**, pp. 371.

109. Zanotelli, M. R., et al. (2016). Stable engineered vascular networks from human induced pluripotent stem cell-derived endothelial cells cultured in synthetic hydrogels. *Acta Biomater*, **35**, pp. 32–41.

110. Kim, S., Lee, H., Chung, M. and Jeon, N. L. (2013). Engineering of functional, perfusable 3D microvascular networks on a chip. *Lab Chip*, **13**, pp. 1489.

111. Kim, S., Chung, M., Ahn, J., Lee, S. and Jeon, N. L. (2016). Interstitial flow regulates the angiogenic response and phenotype of endothelial cells in a 3D culture model. *Lab Chip*, **16**, pp. 4189–4199.

112. Hwang, Y.-S., et al. (2009). Microwell-mediated control of embryoid body size regulates embryonic stem cell fate via differential expression of WNT5a and WNT11. *Proc Natl Acad Sci U S A*, **106**, pp. 16978–16983.

113. Warmflash, A., Sorre, B., Etoc, F., Siggia, E. D. and Brivanlou, A. H. (2014). A method to recapitulate early embryonic spatial patterning in human embryonic stem cells. *Nat Methods*, **11**, pp. 847–854.

114. Nazareth, E. J. P., et al. (2013). High-throughput fingerprinting of human pluripotent stem cell fate responses and lineage bias. *Nat Methods*, **10**, pp. 1225–1231.

115. Smith, Q., Stukalin, E., Kusuma, S., Gerecht, S. and Sun, S. X. (2015). Stochasticity and spatial interaction govern stem cell differentiation dynamics. *Sci Rep*, **5**.

116. Etoc, F., et al. (2016). A balance between secreted inhibitors and edge sensing controls gastruloid self-organization. *Dev Cell*, **39**, pp. 302–315.

117. Martyn, I., Kanno, T. Y., Ruzo, A., Siggia, E. D. and Brivanlou, A. H. (2018). Self-organization of a human organizer by combined Wnt and Nodal signalling. *Nature*, **558**, pp. 132–135.

118. Tewary, M., et al. (2017). A stepwise model of reaction-diffusion and positional information governs selforganized human peri-gastrulation-like patterning. *Development*, **144**, pp. 4298.

119. Kusuma, S., Smith, Q., Facklam, A. and Gerecht, S. (2017). Micropattern size-dependent endothelial differentiation from a human induced pluripotent stem cell line. *J Tissue Eng Regen Med*, **11**(3), pp. 855–861.

120. Ma, Z., et al. (2015). Self-organizing human cardiac microchambers mediated by geometric confinement. *Nat Commun*, **6**, pp. 7413.

121. Smith, Q., et al. (2018). Cytoskeletal tension regulates mesodermal spatial organization and subsequent vascular fate. *Proc Natl Acad Sci U S A*, **115**(32), pp. 8167–8172.

122. Hull, C. W. (1986). Google Patents.

123. Miller, J. S., et al. (2012). Rapid casting of patterned vascular networks for perfusable engineered three-dimensional tissues. *Nat Mater*, **11**, pp. 1–7.

124. Gaetani, R., et al. (2012). Cardiac tissue engineering using tissue printing technology and human cardiac progenitor cells. *Biomaterials*, **33**, pp. 1782–1790.

125. Feinberg, A. W., et al. (2007). Muscular thin films for building actuators and powering devices. *Science*, **317**, pp. 1366.

126. Zhang, Y. S., et al. (2015). From cardiac tissue engineering to heart-on-a-chip: beating challenges. *Biomed Mater*, **10**, pp. 034006.

127. Grosberg, A., Alford, P. W., McCain, M. L. and Parker, K. K. (2011). Ensembles of engineered cardiac tissues for physiological and pharmacological study: heart on a chip. *Lab Chip*, **11**, pp. 4165–4173.

128. Lind, J. U., et al. (2017). Cardiac microphysiological devices with flexible thin-film sensors for higherthroughput drug screening. *Lab Chip*, **17**, pp. 3692–3703.

129. Lind, J. U., et al. (2016). Instrumented cardiac microphysiological devices via multimaterial threedimensional printing. *Nat Mater*, **16**, pp. 303.

130. Liu, Y.-W., et al. (2018). Human embryonic stem cell–derived cardiomyocytes restore function in infarcted hearts of non-human primates. *Nat Biotechnol*, **36**, pp. 597.

131. Vunjak-Novakovic, G., et al. (2009). Challenges in cardiac tissue engineering. *Tissue Eng Part B: Rev*, **16**, pp. 169–187.

132. Burdick, J. A., Mauck, R. L. and Gerecht, S. (2016). To serve and protect: hydrogels to improve stem cell-based therapies. *Cell Stem Cell*, **18**, pp. 13–15.

133. Ronaldson-Bouchard, K., et al. (2018). Advanced maturation of human cardiac tissue grown from pluripotent stem cells. *Nature*, **556**, pp. 239–243.

Chapter 5

Mechanotransduction in Heart Formation

Sandra Rugonyi
Biomedical Engineering, Oregon Health & Science University
3303 SW Bond Ave., Portland, OR 97239, USA
rugonyis@ohsu.edu

This chapter attempts to describe the intricate relation between blood flow dynamics (hemodynamics) and heart development. The heart is one of the first functional organs, and as soon as a primitive tubular heart structure is formed, it starts pumping blood. The tubular heart, which is initially straight, bends and loops, and cardiac looping is followed by heart septation and valve formation. Therefore, important cardiogenesis events occur under blood flow conditions, which in turn influence cardiovascular development. Blood flow is sensed by cardiovascular cells, initiating physical, chemical, and biological responses, which influence the way the heart develops. In essence, blood flow provides mechanical control feedback during heart development, allowing the heart and embryo to adapt to the ever-changing hemodynamic conditions, while ensuring proper cardiac development that optimizes heart function.

Modern Mechanobiology: Convergence of Biomechanics, Development, and Genomics
Edited by Juhyun Lee, Sharon Gerecht, Hanjoong Jo, and Tzung Hsiai

ISBN 978-981-4800-58-7 (Hardcover), 978-0-429-29483-9 (eBook)
www.jennystanford.com

However, when conditions are too far from normal, these same control mechanisms can instead lead to detrimental modifications contributing to heart malformations.

We start this chapter by briefly describing how blood flow affects tissues through mechanotransduction mechanisms. Then we describe normal cardiovascular development and how altered blood flow dynamics could lead to cardiac malformation phenotypes.

5.1 Introduction: Blood Flow Dynamics and Mechanotransduction

Blood flow conditions provide mechanical stimuli to cells. In response to mechanical stimuli, cells and tissues initiate signaling cascades that result in the modulation of different processes, such as secretion of extracellular matrix (ECM) components, cell proliferation, and cell differentiation. In the mature cardiovascular system, this modulation is such that normal (homeostatic) conditions are restored. During cardiac development, however, when hemodynamic conditions are changing over time to meet the demands of the growing embryo, hemodynamic stimuli provide mechanical feedback, which is necessary to ensure the heart develops and functions properly. Flow mechanotransduction, therefore, constitutes a finely tuned feedback loop system that allows tissues to optimally adapt to diverse blood flow environments and respond to deviations from normal conditions.

5.1.1 Mechanical Stimuli in the Cardiovascular System

Mechanical stimuli in the cardiovascular system mainly derive from the interaction between tissues and blood flow. Blood flow exerts wall shear stresses on the cardiovascular surfaces exposed to flow, and blood pressure imposes wall stresses on tissues and cells, regardless of whether they are in direct contact with blood flow (see Fig. 5.1). Application of wall shear stress, a tangential force per unit of surface area, produces a shear deformation as the cell membrane in contact with the flow slightly deforms in the direction of flow. This shear deformation exerts forces on cell-cell junctions as well as cell-ECM attachments and on the cellular cytoskeleton, which also deforms with the cell. Any molecule or subcellular structure, as well as deformation-sensitive ion channels, therefore, could act as a mechanotransducer of blood flow forces. While there can be shear stresses in the interior of the wall that affect other cells, wall shear stresses on endothelial and endocardial cells (ECs), which are in direct contact with blood flow, are directly linked to the local flow characteristics. ECs act as transducers of wall shear stress and, thus, flow. Indeed, ECs typically align in the direction of flow, but the alignment is lost in regions of perturbed flow [1–3].

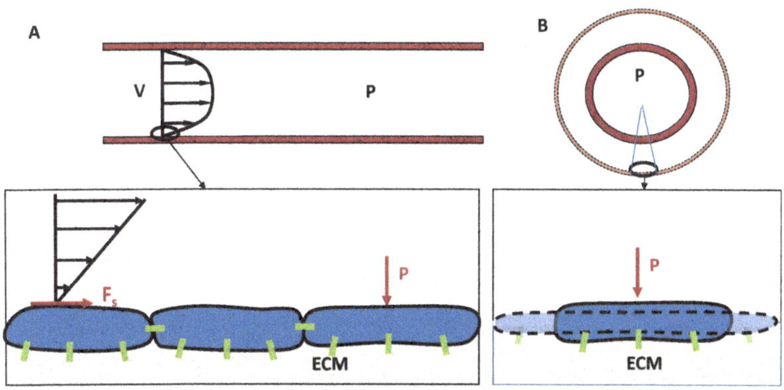

Figure 5.1 Sketch of cardiovascular mechanical stimuli exerted by blood flow on a vessel. (A) Wall shear stress. The flow of blood over endothelial and endocardial cells (ECs) exerts a force tangential to the cell surface. ECs are attached to each other by cell-cell junctions and to the extracellular matrix (ECM). (B) Blood pressure. Changes in blood pressure lead to changes of vessel dimensions (e.g., an increase in blood pressure results in an increase in vessel diameter). Tissue cells are, therefore, stretched in the circumferential direction.

Application of blood pressure deforms all cells in the wall. In a passive blood vessel, increases in blood pressure result in tissue (and cell) stretch, mainly in the circumferential direction. This in turn affects cell-cell junctions, cell-ECM attachments, as well as cytoskeletal organization, which could also act as transducers of flow. Over the course of the cardiac cycle, cells are dynamically stimulated, and they experience cyclic changes in their physical configuration (deformations). Thus, all cells in cardiovascular tissues could respond to the blood flow environment.

Cardiovascular tissues, however, are also exposed to other mechanical stimuli [4, 5]. Typically, the tissues of arteries and vascular segments are stretched longitudinally and circumferentially. This can be tested empirically: if a portion of a vascular segment is excised, its longitudinal dimensions typically shorten, while a longitudinal cut of the excised segment results in a circumferential opening of the tissue (see Fig. 5.2). Therefore, vascular tissues are subjected to residual stresses, which are stresses that are inherent to the tissues that are present even in the absence of flow or blood pressure and modify the tissue material properties by imposing stress conditions even in an unloaded configuration. Residual stresses, however, are very difficult, if not impossible to predict other than empirically.

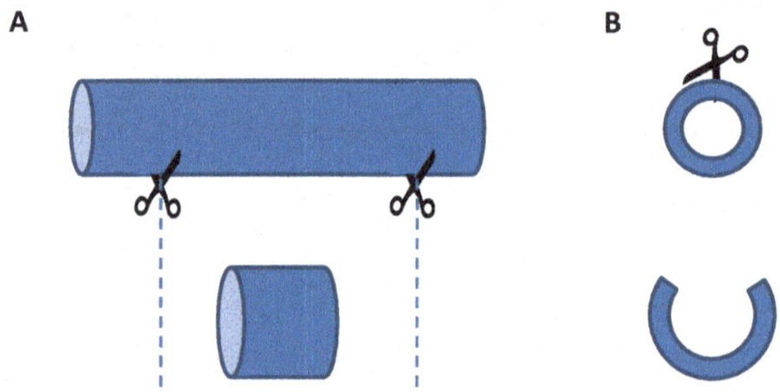

Figure 5.2 Illustration of the effect of residual stresses on cardiovascular vessels. (A) Longitudinal stress: Cutting a portion of a vessel results in longitudinal shortening. (B) Circumferential stress: A longitudinal cut results in circumferential opening of the vessel wall.

In light of the preceding discussion, two important aspects need to be emphasized. (i) It is almost impossible to precisely calculate wall stresses in vivo, especially under pathophysiological conditions, since neither are residual stresses known nor can they be easily inferred. As mentioned before, estimation of residual stresses typically requires invasive access to the tissues, and while this might be possible in certain experiments (e.g., the blood vessel is excised after in vivo measurement) it is frequently not an option for human diagnosis. Residual stresses, further, change under pathophysiological conditions. And (ii) in vivo conditions are very difficult if not impossible to reproduce in vitro. Experiments in vitro frequently decouple the effects of wall shear stress and blood pressure, in the hopes of simplifying the response. For instance, when ECs are seeded on top of a collagen gel or other substrate and flow passes through a chamber containing the seeded cells, effects of wall shear stress can be studied. The cyclical expansion and contraction of the endocardial cell layer during the cardiac cycle, however, is ignored. Similarly, when cells are seeded on elastic structures, which are then stretched, the effects of cell expansion and contraction mainly due to changes in blood pressure can be studied. The effects of wall shear stress, in this case, are neglected. Please note, however, that simply seeding cells and increasing the blood pressure in the media without initiating flow does not really reproduce any in vivo effect: the cells are only pushed against the substrate, but without being able to stretch laterally, as they would in vivo (see Fig. 5.2). Another consideration is that cells are very responsive to their environment: they will respond differently, even to the same mechanical stimuli, depending on the substrate and/or surrounding ECM characteristics. While conducting in vitro experiments, researchers must be aware of these limitations and the many assumptions behind results obtained in this way. In vitro experimentation, however, can be a powerful tool to decouple effects but must be carried out in realistic situations.

5.1.2 Sensing Blood Flow

The stresses and stretches resulting from the interaction of tissues and blood flow are sensed by ECs, smooth muscle cells, fibroblasts, and myocardial cells in several different ways. ECs, which form a monolayer at the interface between cardiovascular tissues and blood, are sensitive to wall shear stresses [2, 3, 6]. However, there is also evidence that ECs respond to stretch [7, 8]. It is widely appreciated that regions of the vasculature with highly oscillatory flows are more prone to atherosclerosis, while regions of slow or stagnant flow are prone to thrombosis and endothelial leakage. While in vivo it is very difficult to separate the effects of shear from those of stretching due to pressure, in vitro experiments (no stretch) show that cells indeed respond to wall shear stress. Smooth muscle cells, as well as myocardial and fibroblast cells, which are not in direct contact with blood, are sensitive to stretches and stresses, which result mainly from the interaction between tissues and blood flow pressure. In general, increased blood pressure over time is associated with thicker and stiffer arterial and cardiac walls.

Several mechanosensors have been proposed, including cilia in ECs [9], integrin signaling components, G-protein receptors, tyrosine kinase receptors, stretch-sensitive ion channels, intracellular junction protein signaling, and membrane lipids [6, 10]. Mechanosensors initiate signaling of mechanotransduction pathways, including Ras, Rho GTPases, mitogen-activated protein (MAP) kinases, Ca^{2+} currents, nitric oxide (NO) signaling, and microRNAs [10]. These pathways then lead to remodeling of cardiovascular tissues (e.g., increased collagen deposition), which leads to physical tissue changes (e.g., increased wall stiffness) [11]. This mechanical feedback loop mechanism, therefore, ensures that cardiovascular tissues can withstand long-term changes in hemodynamic conditions (e.g., prolonged hypertension).

5.1.3 Responses to Blood Flow

Pathophysiological conditions arise, however, when the mechanical stimuli deviate too much from normal or persist over relatively long periods of time. At the onset of many pathophysiological conditions, further deviatory stimuli aggravate the pathophysiology, leading to cardiovascular complications.

In the mature cardiovascular system changes in hemodynamic conditions lead to predictable changes in cardiovascular geometry and tissue composition. Increased blood pressure leads to increased deposition of collagens I and III and wall thickening, which increase tissue stiffness. Increased wall shear stress leads to increased production of NO, a potent vasodilator, and over time increased vessel diameter. The realization that these changes result in the restoration of homeostatic (normal) conditions led to the formulation of a constrained mixture model of growth and remodeling. This model states that growth and remodeling of cardiovascular tissues in response to changes in hemodynamic conditions are such that they restore homeostatic tissue stress levels [4, 11–13]. Thus, with appropriate parameters, deposition of ECM components over time in response to changes in hemodynamics can be predicted [12, 13].

The growth and remodeling theory is easily illustrated as it applies to arteries. Assume a cylindrical, thin-walled vessel in which blood flows. Cyclic changes in blood flow over the cardiac cycle result in cyclic changes in wall stresses. However, to simplify our arguments, let us assume that blood flow has a parabolic velocity profile (Poiseuille flow). In this case, wall shear stress, τ, can be expressed as a function of the vessel's internal radius, R, and volume flow rate, Q, as

$$\tau = (4\mu Q)/(\pi R^3), \tag{5.1}$$

where μ is the blood viscosity. Thus, if we were to keep Q constant, a reduction in τ can be achieved by an increase in R. This is accomplished in the short term by increased release of NO, which dilates the blood vessel, increasing R, followed by more permanent vessel remodeling that adds tissue mass, ensuring an increase in vessel diameter.

Now consider also the equations of stress equilibrium in a thin-walled vessel (also known as Laplace law). Circumferential stress, σ, depends on blood pressure, P, and the vessel thickness, h, as follows:

$$\sigma = PR/h \tag{5.2}$$

Keeping P constant, a reduction of σ can be achieved by increasing the vessel thickness. In cardiovascular vessels this is typically achieved by increased deposition of ECM components, primarily collagens I and III, which increases the tissue volume and thus the vessel thickness. In the constrained mixture model, the rate of deposition of ECM components is proportional to the deviation of σ from homeostatic conditions, allowing prediction of deposition rates. In practice, the different components of the stress tensor (σ_{ij}) can change at similar or different rates, or instead a common stress metric, for example, von Misses stress, can be used to drive deposition. Remodeling changes in response to increased blood pressure, namely addition of collagen to tissues, are typically difficult to revert and thus permanent.

During cardiac development, changes are rapidly taking place, and thus the concept of homeostasis is not really applicable. While it is possible that a more involved form of the growth and remodeling theory applies to cardiac development, such formulation has not yet been developed, and the theory of growth and remodeling does not generally apply to cardiovascular development, except in a few specific cases [4, 14, 15]. For instance, during early tubular heart development, increases in wall shear stress in the heart outflow tract (OFT) lead to an increase in OFT diameter that restores wall shear stress levels to previous levels [16], suggesting that blood flow may induce a shear-mediated vasodilation response in the developing heart OFT.

During stages of heart development, moreover, mechanotransduction signaling affects unique pathways involved in development [10] influencing, among other processes, heart chamber formation, trabeculation, cardiomyocyte proliferation and maturation, and valve formation. While the effects of blood flow on such processes seem to depend on the level of hemodynamic perturbation [17, 18], mathematical models of these processes have not yet been formulated. Development of such models (including experimental validation) could help move forward our understanding of heart formation and how anomalous blood flow can lead to cardiovascular signaling network dysregulation and remodeling, contributing to heart malformations.

5.2 Cardiovascular Development

Cardiac development is an intrinsically multiscale problem. An exquisite orchestration of preprogrammed genetic events prompts cells to proliferate, differentiate, migrate, apoptose, and secrete ECM proteins at the right time and in the right place, giving rise to morphogenetic events that end with heart formation. Genetic programs, however, are modulated by mechanical stimuli. This modulation allows the embryo to survive harsh in utero conditions but can also lead to cardiac malformations. To fully understand the origins of congenital heart disease (CHD), therefore, it is necessary to study the heart and its mechanics as it develops.

5.2.1 Heart Formation

The heart is the first functional organ. Early during embryonic development (about 3 weeks of gestation in humans) a primitive, linear tubular heart is formed. Soon after formation, the tubular heart starts beating and pumping blood by a peristaltic-like mechanism (similar to a peristaltic pump). The heart then bends and twists, forming a looping heart tube. At these initial stages of heart formation, cardiac walls consist of a thin layer of contractile myocardial cells (the myocardium), a monolayer of ECs in contact with blood flow (the endocardium), and an ECM layer in between the myocardium and the endocardium (the cardiac jelly). Further, different cardiac structures start taking shape in the tubular heart. From inlet to outlet, the tubular heart consists of a primitive atrium, an atrioventricular (AV) canal, a ventricle, and an OFT. During tubular heart stages, endocardial cushions, which are precursors of valves and a small portion of the interventricular septum, develop in the AV canal and heart OFT. The cushions are basically non-uniform thickenings of cardiac jelly in the heart wall (typically two opposing cushions) and initially function as primitive valves by closing the lumen upon myocardial contraction. An endothelial-mesenchymal transition (EMT) then takes place, first in the AV cushions and then in the OFT cushions. During EMT, ECs detach from the endocardial layer, acquire a migration phenotype, and start migrating into the cushions. Migrating cells populate the cushions and change the cardiac jelly ECM composition by secreting different proteins as they move and proliferate. Endocardial cushion remodeling, which involves increases in cell density and changes in ECM composition, precedes valve formation and cardiac septation. Thus, appropriate EMT progression is required for proper heart formation. By the end of the first trimester (about 50 days of gestation in humans) the heart is fully formed, but it continues to grow and mature in preparation for birth. This last maturation step involves increase in the number of cells, in particular myocardial cells, and maturation of the myocardial contractile system by careful organization of myofibrils and mitochondria to achieve maximum cardiac efficiency.

5.2.2 Heart Malformation

Congenital cardiac defects occur in about 1% of newborn babies, and they are the leading cause of non-infectious deaths in babies and children in the developed world. CHD is also the most common type of congenital malformation in babies. This is in a way not entirely surprising, given the complex processes that give rise to heart formation.

A number of genes and chromosome mutations have been associated with cardiac defects. For example, a mutation in JAGGED-1 has been associated with tetralogy of Fallot (TOF) [19], as well as deletion of chromosome 22q11.2 [20]; a mutation in NKX2.5 was associated with atrial septal defect [21] [22]; mutations in ZIC3 and TBX20 were associated with double-outlet right ventricle (DORV) [22, 23]; and a mutation in MYH7 was associated with ventricular septal defects (VSDs) [24]. Mutations of several other genes that play important roles in cardiac development have also been linked to CHD, including mechanotransduction genes [21–23, 25, 26].

In addition to gene and chromosomal mutations, exposure to teratogens during gestation has also been associated with cardiac malformations. For example, prenatal exposure to ethanol and retinoic acid has been associated with TOF and DORV [23], while exposure to theophylline, a drug used to treat respiratory diseases, is associated with DORV, transposition of the great arteries and hypoplastic left ventricle [27]. Meanwhile, maternal nutrition can affect heart formation [28], as can maternal diabetes [29, 30], with maternal diabetes predominantly leading to DORV and truncus arteriosus [31].

The preceding discussion shows that it is relatively easy to alter the course of normal heart development, leading to cardiac malformations. Gene mutations and teratogens can certainly alter cardiac development. Perhaps less evident, blood flow can also alter cardiac development. Blood flow is an integral part of heart development, constantly providing mechanical feedback to ensure proper cardiac formation and function. In this way, we can think of the blood flow feedback as a mechanism regulating cardiac development. When mechanical feedback is anomalous, malformations can develop.

5.3 Effect of Blood Flow on Cardiac Formation

Cardiac development is the result of genetic programs modulated by in utero environmental factors [32–35], including biomechanical factors. Abnormal cardiac biomechanics, for example, due to abnormal blood flow conditions caused by placental anomalies or the mother's obesity, result in the formation of heart defects, and thus CHD [36, 37]. The vast majority of CHD cases do not present familial history, and genetic tests are normal [38]. Many researchers now believe that most CHD cases are the result of complex etiologies that include environmental in utero exposure (e.g., smoking, diabetes, and placenta anomalies). Abnormal blood flow conditions during critical embryonic developmental stages are now accepted to be the likely cause of many CHD cases.

The heart pumps blood from its initial tubular stages. Cardiac morphological processes, such as cardiac looping, endocardial cushion formation, EMT, valve formation, and cardiac septation, thus, occur while cardiac tissues are interacting with blood flow dynamics. Blood flow influences any of these morphological processes.

The importance of blood flow in heart development is supported by mutant mice and zebrafish models [23, 26]. Mutations affecting cardiac contractile proteins, which impact blood flow, present defects of endocardial cushions. For example, mouse models with mutations of cardiac troponin T and Na^2 - Ca^2 exchanger (Ncx1) that prevent the heart from beating show endocardial cushion defects and ventricular underdevelopment. Further, mutations affecting known flow mechanotransduction genes (ET-1, NOS3, KLF2, etc.) also show defects associated with abnormal endocardial cushion formation; and mutations in transcription factors (NKX2.5, GATA4, TBX1, TBX5, etc.) lead to cardiac malformations, with similar defects found after altering hemodynamic conditions [6, 19, 23, 26, 39–42]. The details of the mechanisms involved in mechanical feedback during heart formation are only starting to emerge. Different animal models are currently being used to better understand mechanotransduction mechanisms during heart development.

5.3.1 Animal Models of Cardiac Development

Several animal models can be used to study cardiac development. Each model has its own advantages and disadvantages, and therefore different models are used depending on the specific aspect being investigated. Among mammal models, perhaps the most widely used animal model of cardiac development is the mouse. Because genetic manipulations on mouse models are common practice, these models have been extensively used for studying the effects of different genes on cardiac formation, and even for tracing cell lineages [43, 44]. The main disadvantage of mouse models is that it is very difficult to image cardiac development and cardiac function in vivo as non-invasive imaging techniques do not have the resolution and penetration needed. In addition, the placenta of the mouse is very different from that of humans, making the mouse a non-ideal model to study placenta-cardiovascular interactions in the embryos. Guinea pig models, on the other hand, are excellent models for studying the influence of the placenta on embryonic development and heart formation [45, 46]. This is because the guinea pig placenta is very similar to human placentas. As in the case of mouse models, cardiac development in guinea pig embryos is difficult to follow in vivo. Moreover, because guinea pig models are not widely used, and genetic modifications as well as specific antibodies are not generally available, guinea pig models are barely used in cardiac developmental studies. In both guinea pig and mouse models, in addition, measurement of blood flow dynamics during embryonic development is not possible without disturbing the embryo and its environment; and manipulations to alter blood flow conditions, such as surgical interventions, are also challenging. Recently, embryonic cardiac imaging on mouse embryos was measured using optical coherence tomography (OCT), with excellent resolution [47]. OCT imaging required externalization of embryos from the mother, restricting embryo viability over time. Nevertheless, these new developments are promising to study early deviations of blood flow conditions in genetically modified mice.

Other mammalian models used to study cardiac development are sheep and nonhuman primates. Sheep are an excellent model to study late gestation, as the fetus is large enough to instrument its cardiovascular system and allow hemodynamic manipulations [48–51]. Moreover, the effect of drugs and hormones can be studied by injection in the fetus or mother [52]. Due to cost issues, nonhuman primates are mainly used in long-term studies, such as studies to determine influences of obesity and mother nutrition on offspring cardiovascular health [53]. Study of early embryonic development in sheep and primates poses the same difficulties as those faced in the study of early embryonic development in mice models: access to the embryos for in vivo imaging and manipulation is challenging. While large mammalian models are promising in the study of cardiovascular development, the expenses associated with them are large, and thus only a few labs pursue studies in those models.

Avian and zebrafish models of development are all-time favorites in the study of early embryonic cardiac development. This is because the embryos are transparent, enabling high-resolution optical imaging (confocal, light sheet, OCT, etc.) and easily accessible for in vivo imaging and manipulation. Further, since genetic processes are highly conserved among vertebrate species, cardiac developmental processes relevant to human development can be elucidated using avian and zebrafish models. Genetic manipulations are available in zebrafish, allowing scanning of diverse mutations [54–57]. Hemodynamic manipulations in zebrafish embryos, other than through drugs and compounds, are generally very difficult to perform, somewhat restricting the range of hemodynamic studies that can be performed in zebrafish. Moreover, unlike humans, the zebrafish heart has two chambers, restricting the range of heart defects that can be studied. Nevertheless, given the ease of in vivo imaging, fast developmental time, and moderate costs of establishing zebrafish colonies, zebrafish embryos are promising models extensively used in cardiac developmental studies.

Avian models (typically chicken) are also extensively used in cardiac developmental studies. This is due to the ease with which avian embryos can be accessed within the eggs without affecting cardiac growth or function, the ease of embryo manipulation, and cost-effectiveness. Since the circulation is readily accessible, topical application of drugs directly onto the heart or by injection into the circulation is common practice [58]. Further, heart development in humans and chicks is very similar. Like the human heart, the fully formed chicken heart has four chambers and valves, enabling recapitulation of human defects. The avian embryo, moreover, allows for relatively easy monitoring of different parts of its cardiovascular system in vivo data collection. Hemodynamic manipulations, through surgical interventions, are relatively easy to perform in avians and common practice, making the avian model ideal for altering flow patterns at will and thus for studies on the influence of blood flow on cardiac development.

Several surgical interventions have been performed in chicken embryos to alter blood flow conditions at early embryonic developmental stages [36, 59–61]. These interventions mimic changes in hemodynamic conditions due to several factors, such as genetic anomalies and teratogen exposure and also placenta dysfunction and vitelline anomalies, both of which affect embryonic blood flow dynamics (see Fig. 5.3A). Two of the most extensively used interventions are (i) vitelline vein ligation (VVL; Fig. 5.3C), in which blood flow through the right or left vitelline vein, which returns blood to the heart, is blocked, reducing blood flow through the heart and decreasing wall shear stress [61, 62]; and (ii) OFT banding (OTB; Fig. 5.3D), in which a suture is placed and tightened around the heart OFT, increasing blood pressure in the heart and wall shear stress in the OFT [63–65]. OTB results in a range of hemodynamic perturbations that depend on the band tightness or degree of suture constriction of the OFT walls [66, 67]. Band tightness, calculated as the percent difference of the change in maximum OFT diameter before and after banding, can be used as both a measure of OFT constriction and the level of blood flow perturbation.

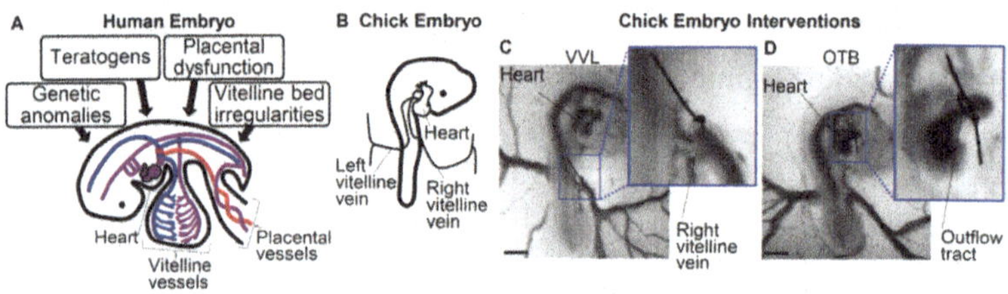

Figure 5.3 Abnormal hemodynamics. (A) Human embryonic circulation with factors contributing to hemodynamic perturbations. (B) Schematic of a normal chick embryo at day 3 (HH18), corresponding to about 4 weeks in human development. Sample optical images of chick embryos at HH18, after surgical interventions used to alter blood flow with (C) vitelline vein ligation (VVL) and (D) outflow tract banding (OTB). Scale bar = 1 mm [68].

Studies that alter normal cardiac blood flow dynamics are elucidating important mechanotransduction processes and mechanisms by which blood flow dynamics influences heart development. Studies aiming at understanding the effects of blood flow dynamics on cardiac development can be generally divided into two groups: (i) studies that focus on early heart changes due to altered blood flow conditions and (ii) studies that focus on the resulting heart malformations later during development, when the heart is fully formed.

5.3.2 Early Embryonic Cardiac Remodeling in Response to Altered Hemodynamics

In vivo manipulations of blood flow conditions typically involve changes in wall shear stress, blood pressure, and tissue motion (stretch). The effects of wall shear stress and wall stretch cannot be isolated in vivo, and the observed biological effect is in response to a combination of mechanical stimuli. Nevertheless, different hemodynamic interventions have been developed to mimic specific changes in blood flow dynamics and sometimes are associated with very specific aspects of blood flow. For example, OTB is frequently associated with increased blood pressure, and biological changes are frequently interpreted in the context of hemodynamic overload; conversely, VVL is frequently associated with a decrease in blood flow rate, with a consequent decrease in wall shear stress and biological changes commonly attributed to altered wall shear stress patterns. Results of in vivo studies should be interpreted with caution in light of the fact that changes in hemodynamics frequently involve changes in both wall shear rates and blood pressure, affecting wall shear rates, wall stresses, and wall stretch at the same time.

5.3.2.1 Effects typically associated with altered wall shear stress

Several EC genes are sensitive to wall shear stress [2, 3]. Among them endothelin-1 (ET1), endothelial nitric oxide synthase (eNOS or NOS3), Krüppel-like factor 2 (KLF2), and Notch have attracted attention in cardiac developmental studies. For example, after VVL intervention in chicken embryos, the expression of ET1, eNOS, and KLF2 with respect to controls significantly changed. In particular, higher expression of eNOS and KLF2 was associated with cardiac regions with increased wall shear stress while higher expression of ET1 was associated with regions of low and disrupted wall shear stress [6]. Regions of high wall shear stress include the surface of endocardial cushions, where KLF2 and eNOS are more highly expressed. After VVL intervention, which reduces blood flow volume through the heart to approximately half its normal levels, eNOS and KLF2 expression in endocardial cushions was increased while ET1 expression in the inner curvature was decreased but increased in the dorsal aorta [6]. These studies showed that ET1, eNOS, and KLF2 are responsive to blood flow dynamics early during development and established a possible role for wall shear mechanotransduction in heart formation and modulation of responses.

Several studies further suggest regulation of EMT rates by hemodynamics [17, 69–73], although the mechanisms remain to be elucidated. These studies involved both in vivo and cushion explant (in vitro) studies. Interestingly, ECs on the cushions, which undergo EMT, do not have cilia, whereas ECs in regions of low shear stress do have cilia to sense wall shear [9]. This led to the speculation that the absence of cilia may be linked to the ability of ECs to undergo EMT [6, 9, 74]. In any case, the rate of EMT is determined by the availability of cells in the endocardium, the rate at which ECs get activated and successfully detach from the endocardium, and the rate of invasion of EMT cells into the cushions [75–77]. It is unclear, however, which EMT factor contributes to variation in EMT rates in response to blood flow conditions deviating from normal. Studies on zebrafish embryos revealed that blood flow affects AV cushion endocardial cell surface density prior to EMT, increasing the overall availability of ECs in the cushion surface, and that KLF2 is involved in flow-induced changes in AV EMT [78]. These data suggest that KLF2 mediates flow-induced changes in EMT in the endocardial cushions.

Studies on chicken embryos further suggest that the rate of EMT in OFT cushions depends on the level of hemodynamic perturbation [17]. After performing OTB interventions, OFT heart tissues were fixed and analyzed. It was found that the density of cells close to the endocardium in endocardial cushions increased in banded hearts compared to control hearts and that this increase depended on band tightness and, therefore, on the level of hemodynamic alteration. These findings suggest that the rate of EMT depends on how much blood flow was altered early during development. This is important for valve formation, as EMT is a critical process that ultimately determines the fate of valve structures.

5.3.2.2 Effects typically associated with altered blood pressure

When the heart has a tubular structure with no valves or chambers (gestation of about 3 to 7 weeks in humans), hemodynamic loading (blood pressure) affects the morphological development of the heart ventricle. As the tubular heart grows, a sponge-like tissue, known as the trabecular myocardium, extends from the compact myocardium layer toward the lumen. Morphological studies of the effects of hemodynamic conditions on ventricular development are widely available (see Refs. [59, 79, 80] for examples), and it appears that growth of trabecular structures is entirely regulated by hemodynamic conditions. Elevated hemodynamic loading leads to excessive trabeculation and later to noncompaction cardiomyopathy; reduced loading leads to less trabeculation. Lack of cardiac trabeculation in embryos leads to severely compromised cardiac function [81] and embryolethality [82]. Abnormal trabeculation, in turn, affects the development of the interventricular septum and results in impaired cardiac function. Infants born with ventricular hypotrabeculation show compromised hemodynamics, and those with ventricular hypertrabeculation (or noncompaction cardiomyopathy) show impaired diastolic function [83]. Noncompaction cardiomyopathies have recently come to attention due to advances in imaging technologies that allow better diagnosis. Depending on the degree of trabecular noncompaction, patient symptoms range from nonsymptomatic to disabling congestive heart failure, arrhythmias, and systemic thromboemboli [84].

Despite the importance of trabeculation on cardiac function—during development and beyond—the role of trabecular structures and the mechanical mechanisms that regulate trabecular formation are virtually unstudied. While probably several genes and biochemical regulatory pathways are at play, it is frequently hypothesized that signaling is initiated by mechanical stimuli (stress or strain) acting on regulatory pathways through mechanotransduction mechanisms. Studies to understand the regulation of blood flow on ventricular trabeculation are underway using zebrafish models [57]. In these studies, genetic manipulations together with morpholino interventions are used to alter blood flow conditions (including blood viscosity) and cardiac morphology and to test the effects of genes on trabeculation. In particular, a focus of those studies is on the modulation of trabeculation by Notch signaling, which is involved in ventricular trabeculation while it is sensitive to wall shear rates. More research is clearly needed in this exciting area to elucidate not only the genes that contribute to ventricular trabeculation but also the hemodynamic stimuli (stretch, shear, or both) that modulate trabecular formation.

5.3.3 Cardiac Malformation Phenotypes after Hemodynamic Interventions

Several studies have determined cardiac malformation phenotypes that emerge after hemodynamic interventions. Investigations on valve formation have mainly focused on the AV valves and used zebrafish models as well as avian models [69, 73, 78, 85, 86]. For other cardiac malformations, chicken embryos have typically been used, given the similarity of the chick heart with the human heart (four chambers), which allows easy identification of human phenotypes in the chick.

Early studies reported a spectrum of cardiovascular malformations after interventions that altered blood flow in chicken [36, 41, 61, 80]. Distinct hemodynamic interventions were characterized and malformation phenotypes assessed. These studies typically involved performing interventions during the tubular heart period and then waiting until the heart was fully formed (but the embryos had not hatched) to determine heart defect phenotypes. Different techniques can be used to assess malformations, such as a combination of ultrasound imaging to first determine cardiac function and then computed tomography (CT) imaging with micron resolution (microCT) to more carefully characterize cardiac malformations (e.g., see Fig. 5.4). The most common malformations found were VSDs, which are also the most common form of heart malformation in humans, but more complex cardiac defects, such as TOF and DORV, also resulted from hemodynamic interventions [41]. These studies established that cardiac malformations found in human babies can be reproduced by altering hemodynamic conditions during early embryonic development. These results opened the possibility that many human cases of CHD could be related to exposure to altered blood flow. Furthermore, because other causes of CHD (gene mutations and teratogen exposure) can also lead to altered hemodynamics, altered blood flow can be an underlying mechanism leading to CHD.

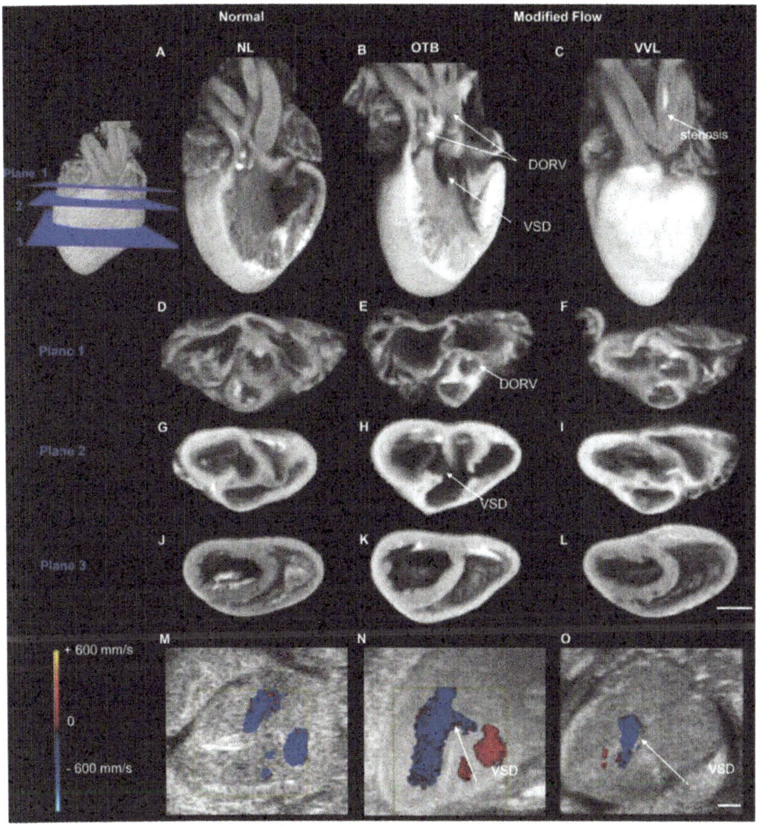

Figure 5.4 Cardiac defects at HH38. Sample microCT images of a normal (A, D, G, J), banded (B, E, H, K), and vein ligated (C, F, I, L) embryo, depicting 3D reconstructions (A–C) and three cross-sectional planes (D–L). Plane 1 intersects the semilunar valves (D–F), plane 2 intersects the atrioventricular valves (G–I), and plane 3 intersects the ventricle midpoint (J–L). DORV in the banded embryo displayed with aortic valve rotated outward (E), along with both outflows from the RV and a perimembranous VSD (B, H). The vein ligated embryo displayed stenosis of the right brachiocephalic artery (C). Sample ultrasound color doppler images (M–O) with detection of VSD flow after both interventions (N, O). Scale bars = 1 mm. NL, normal; VVL, vitelline vein ligated; OTB, outflow tract banded; DORV, double-outlet right ventricle; VSD, ventricular septal defect [68].

A question that remained unanswered, however, was whether different levels of hemodynamic alteration would lead to distinct phenotypes and/or incidence of cardiac malformations. By carefully measuring the alterations in hemodynamic conditions after OTB and VVL interventions and dividing embryos according to different levels of blood flow perturbation (VVL and different ranges of band tightness), cardiac malformation incidence and phenotype were shown to depend on the degree to which blood flow conditions were perturbed with respect to normal or control conditions [68]. Even hearts that were seemingly structurally normal after hemodynamic interventions had minor differences with respect to control hearts. These findings suggest that embryonic exposure to altered blood flow leads to modifications that are detrimental to heart formation. More studies are required, however, to determine the short- and long-term consequences of early exposure to an altered hemodynamic load on cardiac development.

5.4 Conclusions

The fact that specific cardiac defects can be the result of genetic anomalies, teratogen toxicity, and abnormal blood flow independently, [23, 26] is an argument in favor of common pathways leading to specific phenotypes. Altered blood flow patterns may be a common link among different exposures and provide a means for malformations to develop.

Blood flow can alter cardiac formation by diverse mechanisms. Several mechanosensors and signaling cascades are affected by blood flow conditions, and disruptions in any of these can lead to defects. Moreover, during cardiogenesis, signaling as well as physical and biochemical processes act together. Thus, physical and biochemical disruptions have to be also considered when analyzing the effects of blood flow on cardiac development. More research is needed to elucidate the fascinating interaction between flow and cardiogenesis.

References

1. Boon, R. A., et al. (2010). KLF2-induced actin shear fibers control both alignment to flow and JNK signaling in vascular endothelium. *Blood*, **115**(12), pp. 2533–2542.

2. Davies, P. F. (1995). Flow-mediated endothelial mechanotransduction. *Physiol Rev*, **75**(3), pp. 519–60.
3. Davies, P. F. (2009). Hemodynamic shear stress and the endothelium in cardiovascular pathophysiology. *Nat Clin Pract Cardiovasc Med*, **6**(1), pp. 16–26.
4. Taber, L. A. and Humphrey, J. D. (2001). Stress-modulated growth, residual stress, and vascular heterogeneity. *J Biomech Eng*, **123**, pp. 528–535.
5. Fung, Y. C. (1993). *Biomechanics: Mechanical Properties of Living Tissues*, 2nd ed., Springer-Verlag, New York.
6. Groenendijk, B. C., et al. (2007). The role of shear stress on ET-1, KLF2, and NOS-3 expression in the developing cardiovascular system of chicken embryos in a venous ligation model. *Physiology*, **22**, pp. 380–389.
7. Jufri, N. F., et al. (2015). Mechanical stretch: physiological and pathological implications for human vascular endothelial cells. *Vasc Cell*, **7**, pp. 8.
8. Helmke, B. P., Rosen, A. B. and Davies, P. F. (2003). Mapping mechanical strain of an endogenous cytoskeletal network in living endothelial cells. *Biophys J*, **84**(4), pp. 2691–2699.
9. Van der Heiden, K., et al. (2006). Monocilia on chicken embryonic endocardium in low shear stress areas. *Dev Dyn*, **235**, pp. 19–28.
10. Andrés-Delgado, L. and Mercader, N. (2016). Interplay between cardiac function and heart development. *Biochim Biophys Acta*, **1863**(7, Part B), pp. 1707–1716.
11. Humphrey, J. D. (2001). Stress, strain, and mechanotransduction in cells. *J Biomech Eng*, **123**, pp. 638–641.
12. Humphrey, J. D. and Rajagopal, K. R. (2003). A constrained mixture model for arterial adaptations to a sustained step change in blood flow. *Biomech Model Mechanobiol*, **2**, pp. 109–126.
13. Valentin, A. and Humphrey, J. D. (2009). Evaluation of fundamental hypotheses underlying constrained mixture models of arterial growth and remodeling. *Philos Trans R Soc A*, **367**, pp. 3585–3606.
14. Taber, L. A. (1998). An optimization principle for vascular radius including the effects of smooth muscle tone. *Biophys J*, **74**, pp. 109–114.
15. Lee, L. C., Kassab, G. S. and Guccione, J. M. (2016). Mathematical Modeling of Cardiac Growth and Remodeling. *Wiley Interdiscip Rev: Syst Biol Med*, **8**(3), pp. 211–226.

16. Midgett, M., et al. (2015). Blood flow through the embryonic heart outflow tract during cardiac looping in HH13–HH18 chicken embryos. *J R Soc Interface*, **12**(111).

17. Midgett, M., et al. (2017). Increased hemodynamic load in early embryonic stages alters endocardial to mesenchymal transition. *Front Physiol*, **8**(56).

18. Midgett, M., et al. (2017). Increased hemodynamic load in early embryonic stages alters myofibril and mitochondrial organization in the myocardium. *Front Physiol*, **8**(631).

19. Eldadah, Z. A., et al. (2001). Familial Tetralogy of Fallot caused by mutation in the jagged1 gene. *Hum Mol Genet*, **10**(2), pp. 163–169.

20. Momma, K., et al. (2001). Tetralogy of Fallot associated with chromosome 22q11.2 deletion in adolescents and young adults. *Genet Med*, **3**(1), pp. 56–60.

21. Srivastava, D. and Olson, E. N. (2000). A genetic blueprint for cardiac development. *Nature*, **407**(6801), pp. 221–226.

22. Finsterer, J., Stollberger, C. and Towbin, J. A. (2017). Left ventricular noncompaction cardiomyopathy: cardiac, neuromuscular, and genetic factors. *Nat Rev Cardiol*, **14**(4), pp. 224–237.

23. Rugonyi, S. (2016). Genetic and flow anomalies in congenital heart disease. *AIMS Genet*, **3**(3), pp. 157–166.

24. Bettinelli, A. L., et al. (2013). Familial ebstein anomaly, left ventricular hypertrabeculation, and ventricular septal defect associated with a MYH7 mutation. *Am J Med Genet Part A*, **161**(12), pp. 3187–3190.

25. Bruneau, B. G. (2003). The developing heart and congenital heart defects: a make or break situation. *Clin Genet*, **63**(4), pp. 252–261.

26. Granados-Riveron, J. T. and Brook, J. D. (2012). The impact of mechanical forces in heart morphogenesis. *Circulation*, **5**(1), pp. 132–142.

27. Park, J. M., Schmer, V. and Myers, T. L. (1990). Cardiovascular anomalies associated with prenatal exposure to theophylline. *South Med J*, **83**(12), pp. 1487–1488.

28. Mills, J. L., et al. (2010). Maternal obesity and congenital heart defects: a population-based study. *Am J Clin Nutr*, **91**(6), pp. 1543–1549.

29. Rowland, T. W., Hubbell, Jr., J. P. and Nadas, A. S. (1973). Congenital heart disease in infants of diabetic mothers. *J Pediatr*, **83**(5), pp. 815–820.

30. Wren, C., Birrell, G. and Hawthorne, G.(2003). Cardiovascular malformations in infants of diabetic mothers. Heart, **89**(10), pp. 1217–1220.

31. Ferencz, C., et al. (1990). Maternal diabetes and cardiovascular malformations: Predominance of double outlet right ventricle and truncus arteriosus. *Teratology*, **41**(3), pp. 319–326.

32. Barker, D. J., et al. (2010). The early origins of chronic heart failure: impaired placental growth and initiation of insulin resistance in childhood. *Eur J Heart Fail*, **12**, pp. 819–825.

33. Barker, D. J., et al. (1989). Growth in utero, blood pressure in childhood and adult life, and mortality from cardiovascular disease. *Br Med J*, **298**, pp. 564–567.

34. Thornburg, K. L. (2011). Foetal programming reveals the dark side of AT(2)R. *Cardiovasc Res*, **89**, pp. 260–261.

35. Thornburg, K. L., O'Tierney, P. F. and Louey, S. (2010). The placenta is a programming agent for cardiovascular disease. *Placenta*, (Suppl), pp. S54–S59.

36. Clark, E. B. and Rosenquist, G. C. (1978). Spectrum of cardiovascular anomalies following cardiac loop constriction in the chick embryo. *Birth Defects: Original Article Series*, **XIV**(7), pp. 431–442.

37. Hove, J. R., et al. (2003). Intracardiac fluid forces are an essential epigenetic factor for embryonic cardiogenesis. *Nature*, **421**, pp. 172–177.

38. Øyen, N., et al. (2009). Recurrence of congenital heart defects in families. *Circulation*, **120**, pp. 295–301.

39. Obler, D., et al. (2008). Double outlet right ventricle: aetiologies and associations. *J Med Genet*, **45**(8), pp. 481–497.

40. Fujita, M., et al. (2016). Pharyngeal arch artery defects and lethal malformations of the aortic arch and its branches in mice deficient for the Hrt1/Hey1 transcription factor. *Mech Dev*, **139**, pp. 65–73.

41. Midgett, M. and Rugonyi, S. (2014). Congenital heart malformations induced by hemodynamic altering surgical interventions. *Front Physiol*, **5**.

42. Bruneau, B. G. (2008). The developmental genetics of congenital heart disease. *Nature*, **451**(7181), pp. 943–948.

43. Moon, A. (2008). Mouse models of congenital cardiovascular disease, in *Current Topics in Developmental Biology*. Academic Press, pp. 171–248.

44. Meilhac, S. M., et al. (2014). Cardiac cell lineages that form the heart. *Cold Spring Harbor Perspect Med*, **4**(9), pp. a013888.

45. Mess, A. (2007). The guinea pig placenta: model of placental growth dynamics. *Placenta*, **28**(8), pp. 812–815.

46. Saiki, Y., et al. (1997). Hemodynamic alteration by fetal surgery accelerates myocyte proliferation in fetal guinea pig hearts. *Surgery*, **122**(2), pp. 412–419.

47. Syed, S. H., et al. (2011). Optical coherence tomography for high-resolution imaging of mouse development in utero. *J Biomed Opt*, **16**(4), pp. 046004.

48. Reller, M. D., et al. (1992). Severe right ventricular pressure loading in fetal sheep augments global myocardial blood flow to submaximal levels. *Circulation*, **86**, pp. 581–588.

49. Barbera, A., et al. (2000). Right ventricular systolic pressure load alters myocyte maturation in fetal sheep. *Am J Physiol Heart Circ Physiol*, **279**, pp. R1157–R1164.

50. Jonker, S. S., et al. (2006). Sequential growth of fetal sheep cardiac myocytes in response to simultaneous arterial and venous hypertension. *Am J Physiol Heart Circ Physiol*, **292**, pp. R913– R919.

51. Thornburg, K., et al. (2011). Regulation of the cardiomyocyte population in the developing heart. *Prog Biophys Mol Biol*, **106**(1), pp. 289–299.

52. Chattergoon, N. N., et al. (2012). Thyroid hormone drives fetal cardiomyocyte maturation. *FASEB J*, **26**(1), pp. 397–408.

53. Muralimanoharan, S., et al. (2017). Sexual dimorphism in the fetal cardiac response to maternal nutrient restriction. *J Mol Cell Cardiol*, **108**, pp. 181–193.

54. Bakkers, J. (2011). Zebrafish as a model to study cardiac development and human cardiac disease. *Cardiovasc Res*, **91**(2), pp. 279–288.

55. Brown, D., et al. (2016). Advances in the study of heart development and disease using zebrafish. *J Cardiovasc Dev Dis*, **3**(2), pp. 13.

56. Lee, J., et al. (2014). hemodynamics and ventricular function in a zebrafish model of injury and repair. *Zebrafish*, **11**(5), pp. 447–454.

57. Lee, J., et al. (2016). 4-Dimensional light-sheet microscopy to elucidate shear stress modulation of cardiac trabeculation. *J Clin Invest*, **126**(5), pp. 1679–1690.

58. Burggren, W. W., Santin, J. F. and Antich, M. R. (2016). Cardio-respiratory development in bird embryos: new insights from a venerable animal model. *Rev Bras Zootecn*, **45**, pp. 709–728.

59. Keller, B. B. (1998). Embryonic cardiovascular function, coupling and maturation: a species view, in *Development of Cardiovascular Systems*, Burggren, W. W. and Keller, B. B., eds. University Press, Cambridge, MA.

60. Lucitti, J. L., Tobita, K. and Keller, B. B. (2005). Arterial hemodynamics and mechanical properties after circulatory intervention in the chick embryo. *J Exp Biol*, **208**, pp. 1877–1885.

61. Hogers, B., et al. (1997). Unilateral vitelline vein ligation alters intracardiac blood flow patterns and morphogenesis in the chick embryo. *Circ Res*, **80**(4), pp. 473–481.

62. Hogers, B., et al. (1999). Extraembryonic venous obstructions lead to cardiovascular malformations and can be embryolethal. *Cardiovasc Res*, **41**(1), pp. 87–99.

63. Clark, E. B., et al. (1989). Effect of increased pressure on ventricular growth in stage 21 chick embryos. *Am J Physiol Heart Circ Physiol*, **257**(1), pp. H55–H61.

64. Taber, L., Keller, B. and Clark, E. (1992). Cardiac mechanics in the stage 16 chick embryo. *J Biomech Eng*, **114**, pp. 427–434.

65. Tobita, K., et al. (2005). Three-dimensional myofiber architecture of the embryonic left ventricle during normal development and altered mechanical loads. *Anat Rec A: Discov Mol Cell Evol Biol*, **283A**(1), pp. 193–201.

66. Shi, L., et al. (2013). Alterations in pulse wave propagation reflect the degree of outflow tract banding in HH18 chicken embryos. *Am J Physiol Heart Circ Physiol*, **305**, pp. H386–H396.

67. Midgett, M., Goenezen, S. and Rugonyi, S. (2014). Blood flow dynamics reflect degree of outflow tract banding in Hamburger-Hamilton stage 18 chicken embryos. *J R Soc Interface*, **11**(100), pp. 20140643.

68. Midgett, M., Thornburg, K. L. and Rugonyi, S. (2017). Blood flow patterns underlie developmental heart defects. *Am J Physiol Heart Circ Physiol*, **312**, H632–H642.

69. Menon, V., et al. (2015). Altered hemodynamics in the embryonic heart affects outflow valve development. *J Cardiovasc Dev Dis*, **2**(2), pp. 108.

70. Goodwin, R. L., et al. (2014). The impact of flow-induced forces on the morphogenesis of the outflow tract. *Front Physiol*, **5**.

71. Goodwin, R. L., et al. (2005). A three-dimensional model system of valvulogenesis. *Dev Dyn*, **233**, pp. 122–129.

72. Potts, J. D., Yost, M. J. and Goodwin, R. L. (2006). Models of cardiovascular development: New approaches are making in vitro en vogue. *Curr Cardiol Rev*, **2**, pp. 55–63.

73. Tan, H., et al. (2013). Fluid flow forces and rhoA regulate fibrous development of the atrioventricular valves. *Dev Biol*, **374**(2), pp. 345–356.

74. Poelmann, R. E., Gittenberger-de Groot, A. C. and Hierck, B. P. (2008). The development of the heart and microcirculation: role of shear stress. *Med Biol Eng Comput*, **46**, pp. 479–484.

75. Mercado-Pimentel, M. E., Hubbard, A. D. and Runyan, R. B. (2007). Endoglin and Alk5 regulate epithelial–mesenchymal transformation during cardiac valve formation. *Dev Biol*, **304**(1), pp. 420–432.

76. Lencinas, A., et al. (2013). Olfactomedin-1 activity identifies a cell invasion checkpoint during epithelial-mesenchymal transition in the chick embryonic heart. *Dis Models Mech*, **6**(3), pp. 632–642.

77. Tavares, A. L. P., et al. (2017). Runx2-I is an early regulator of epithelial-mesenchymal cell transition in the chick embryo. *Dev Dyn*, **247**, pp. 542–554.

78. Steed, E., et al. (2016). klf2a couples mechanotransduction and zebrafish valve morphogenesis through fibronectin synthesis. *Nat Commun*, **7**, pp. 11646.

79. Sedmera, D., et al. (1997). Developmental changes in the myocardial architecture of the chick. *Anat Rec*, **248**, pp. 421–432.

80. Sedmera, D., et al. (1999). Remodeling of chick embryoniv ventricular myoarchitecture under experimentally changed loading conditions. *Anat Rec*, **254**, pp. 238–252.

81. Liu, J., et al. (2010). A dual role for ErbB2 signaling in cardiac trabeculation. *Development*, **137**, pp. 3867–3875.

82. Jenni, R., Rojas, J. and Oechslin, E. (1999). Isolated noncompaction of the myocardium. *N Engl J Med*, **340**, pp. 966–967.

83. Peshkovsky, C., Totong, R. and Yelon, D. (2011). Dependence of cardiac trabeculation on Neuregulin signaling and blood flow in zebrafish. *Dev Dyn*, **240**, pp. 446–456.

84. Weiford, B. C., Subbarao, V. D. and Mulherm, K. M. (2004). Noncompaction of the ventricular myocardium. *Circulation*, **109**, pp. 2965–2971.

85. Lee, Y. M., et al. (2006). Vascular endothelial growth factor receptor signaling is required for cardiac valve formation in zebrafish. *Dev Dyn*, **235**, pp. 29–37.

86. Tan, H., et al. (2011). Expression and deposition of fibrous extracellular matrix proteins in cardiac valves during chick development. *Microsc Microanal*, **17**, pp. 91–100.

Chapter 6

Mechanotransduction in Cardiovascular Development and Regeneration: A Genetic Zebrafish Model

Rongsong Li, Kyung In Baek, Chih-Chiang Chang, Bill Zhou, and Tzung Hsiai
Department of Medicine and Bioengineering, UCLA, 10833 Le Conte Ave, Los Angeles, CA 90095, USA
thsiai@mednet.ucla.edu

Cardiovascular diseases such as coronary heart disease, myocardial infarction (MI), and cardiac arrhythmia are the leading causes of morbidity and mortality in the developed countries and steadily increasing in the developing countries. Fundamental mechanistic studies at the molecular/cellular levels and with animal models are critical for the diagnosis of the diseases and discovery of the related drugs. Despite being phylogenetically distant from humans, zebrafish have remarkable similarity in terms of genetics and electrophysiology of the cardiovascular systems. In the last two decades, the development and deployment of innovative genetic

Modern Mechanobiology: Convergence of Biomechanics, Development, and Genomics
Edited by Juhyun Lee, Sharon Gerecht, Hanjoong Jo, and Tzung Hsiai

ISBN 978-981-4800-58-7 (Hardcover), 978-0-429-29483-9 (eBook)
www.jennystanford.com

manipulation techniques have greatly facilitated the application of zebrafish as models for basic biology and diseases. The critical mechanosensitive signaling pathways in cardiovascular development and pathophysiology previously studied in mammals have been recapitulated in zebrafish. In this chapter, we review the importance of some mechanotransduction pathways, such as Notch, PKCε/PFKFB3, and Wnt/Ang-2, in cardiovascular development and homeostasis and their implications in vascular regeneration/repair in the zebrafish model.

6.1 Introduction of Zebrafish as a Cardiovascular Model

Cardiovascular diseases are the leading cause of death, and their burden is steadily increasing worldwide [1]. While the advancement in biomedical research promises newer, faster, and cheaper diagnoses as well as therapies, the fundamentals of discovery remain deeply rooted in elucidating the pathogenic mechanisms at the molecular and cellular levels, especially in the advent of personalized medicine. By using animal models, gene functions and signal pathways can be linked to normal development and pathophysiology, allowing for the identification and validation of pharmacological targets for novel therapeutics.

The zebrafish has been utilized as an important developmental model because it shares a largely conserved physiology and anatomy with mammals. Its unique transparency at the larval stage allows for direct observation of organ development, including that of the cardiovascular system [2–5]. More recently the role of the zebrafish as a human disease model organism has been expanded to include the adult zebrafish due to a genome that bears similarity to human [4, 6, 7] and the relative ease of genetic manipulation, including the latest genome engineering approaches [8–10]. Simple husbandry and rapid development enable a range of large-scale phenotypic screening [3, 4, 7]. Furthermore, zebrafish demonstrate an impressive regenerative capacity, such as vascular repair and heart regeneration, that scientists hope to unlock in humans for therapeutic benefits.

6.2 ECG in Zebrafish

Zebrafish are an ideal animal model to study cardiovascular development and pathophysiology because zebrafish hearts share common structures with and have electrophysiology similar to mammalian ones [11]. The remarkable regenerative capacity of zebrafish hearts also makes them an excellent genetic model to study the mechanisms of heart regeneration. Zebrafish hearts achieve full regeneration even after a 20% ventricular resection, without either scarring or arrhythmia, whereas human MI results in irreversible loss of cardiomyocytes [12, 13].

Besides its clinical significance, monitoring of electrocardiography (ECG) enables uncovering the underlying mechanotransduction pathways of cardiac diseases for the development of therapeutic and diagnostic tools [14, 15] in basic research. Zebrafish reveal atrial and ventricular electrical signals similar to those found in a human ECG [16, 17]. Thus, zebrafish have been applied to many cardiac physiological studies, genetic and molecular characterizations, and drug screenings using different types of microsensors. We have developed a microsensor prototype for real-time monitoring of heart regeneration and functions under different perturbations as a fundamental model for cardiac diagnosis and therapeutics [18–22].

The first generation of the microsensor to measure ECG in zebrafish was based on two 29-gauge microelectrodes (AD Instrument, Colorado Springs, CO) positioned at 90º to the animal's ventral epidermis [23]. Adult zebrafish were sedated and placed on a damp sponge to expose the ventral side for visualization under the microscope [21]. The second-generation prototype reduced data acquisition time and zebrafish incision with implantable and flexible multimicroelectrode membrane arrays that measure the epicardial electrocardiogram signals of zebrafish in real time. These interfacing flexible microelectrodes provide enhanced spatial resolution at approximately 5 μm for real-time longitudinal monitoring of electrical signals from the nonplanar and dynamic cardiac surface [19]. The third-generation device was a wearable zebrafish harness with a percutaneous planar electrode array, which enabled studying the electrical conduction phenotypes during zebrafish heart regeneration. The wearable harness and electrode array remained functional in the freely swimming zebrafish for a prolonged period of time. This innovation paved the way for future long-term underwater wireless ECG recording in real time [18, 22]. The entire ECG recording process is performed in a Faraday cage to shield against interference from electromagnetic radiation, and wavelet transform was applied to obtain denoised ECG signals [20].

Despite its small signal intensity, zebrafish ECG measured with microsensors revealed a distinguishable P wave (atrial contraction), QRS complexes (ventricular contraction), RR interval (heartbeat), and ST segment (ventricular repolarization) after signal processing, which is remarkably similar to human ECG (Fig. 6.1A) [19]. Sun et al. applied microsensors to study after ventricular amputation in zebrafish and demonstrated a modification of the QTc interval [23]. Yu et al. further improved the technique to demonstrate a prolonged QT interval in regenerated zebrafish hearts despite complete structural regeneration as well as prolonged QRS interval in zebrafish treated with the antiarrhythmic amiodarone [19] (Figs. 6.1B and 6.1C). Thus, the zebrafish not only allows for the assessment of conduction phenotypes with respect to genetic, epigenetic, and pharmacologic perturbations when studying contractility and regeneration but also allows for the study of drug sensitivity and efficacy.

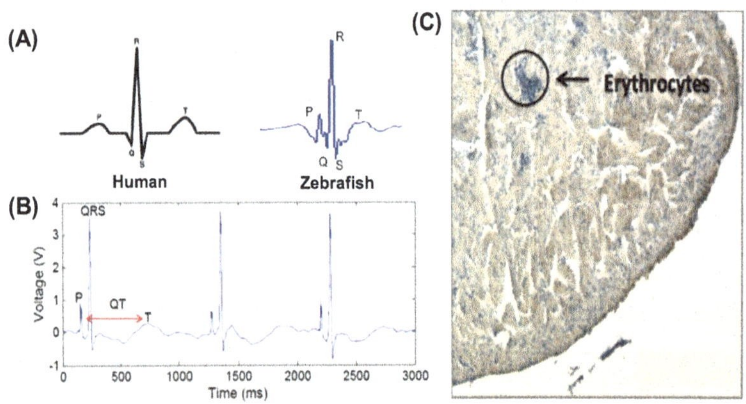

Figure 6.1 Zebrafish ECG with microsensors. (A) Zebrafish ECG signals obtained from microsensors demonstrated high similarity with those of humans. (B) ECG at 59 days after anterior resection of zebrafish heart showed prolonged QTc intervals. (C) Connexin-43 (CX43) staining at 59 days after resection revealed complete structural regeneration of zebrafish heart while QTc intervals were persistently prolonged. Reprinted by permission from Springer Nature Customer Service Centre GmbH: Springer Nature, *Annals of Biomedical Engineering*, Ref. [20], Copyright (2010).

The application of different microsensors enabled the acquisition of the electrical signal of zebrafish hearts with high spatial and temporal resolution. The use of a microelectrode array allowed a non-invasive and longitudinal approach to identify the specific electrical responses to tissue injury, drug screening, and the manipulation of mechanotransduction signaling pathways. The combination of the two has a great potential as an accurate platform to elucidate the mechanism of heart regeneration and provide a minimally invasive tool to monitor and diagnose diseases such as arrhythmia and myocardial infarction.

6.3 Mechanosensitive Pathways Modulate Vascular Development and Regeneration in Zebrafish

Hemodynamic fluid shear stress provides biomechanical cues for the differentiation of stem cells [24, 25] and mesenchymal progenitors [26] into vascular endothelial cells (ECs) [27–29]. Taking advantage of the zebrafish model system, our lab and others have demonstrated that shear stress activates mechanotransduction pathways such as Notch, PKCε/6-phosphofructo-2-kinase/fructose-2,6-biphosphatase 3 (PFKFB3), and Wnt/angiopoietin (Ang)-2 that are implicated in vascular development and regeneration (Fig. 6.2).

6.3.1 Notch Signaling in Vascular Regeneration

The Notch signaling pathway involves a series of cell-fate determination and regulates the initiation of angiogenic sprouting [30, 31]. Upon ligand binding, Notch receptors undergo proteolytic cleavage to release the Notch intracellular cytoplasmic domain (NICD) under the regulation of enzymes in the disintegrin and metalloproteinases (ADAM) family. Following translocation to the nucleus, NICD forms a transcriptional activation complex consisting of recombination signal-binding protein for the immunoglobulin J region (Rbp-Jκ, also known as CSL, CBF1, suppressor of hairless, or Lag-1), mastermind-like (MAML), and forkhead box O subfamily 1 (FOXO1) (FKHR) protein to induce downstream Notch target genes, including hairy and enhancer of split-1 (*Hes1*) and the *gridlock* [32].

Ablation of Notch1 causes developmental retardation, resulting in embryonic lethality [33]. Missense mutation in the Notch3 gene causes the development of the degenerative vascular disease known as cerebral autosomal-dominant arteriopathy with subcortical infarcts and leukoencephalopathy (CADASIL) [34]. Dysregulated Notch activity induces failure of vascular specification and results in abnormal endothelial proliferation, leading to the formation of a hyperplastic vascular network [33, 35]. Notch signaling is further implicated in stem cell differentiation and proliferation [32, 36–39].

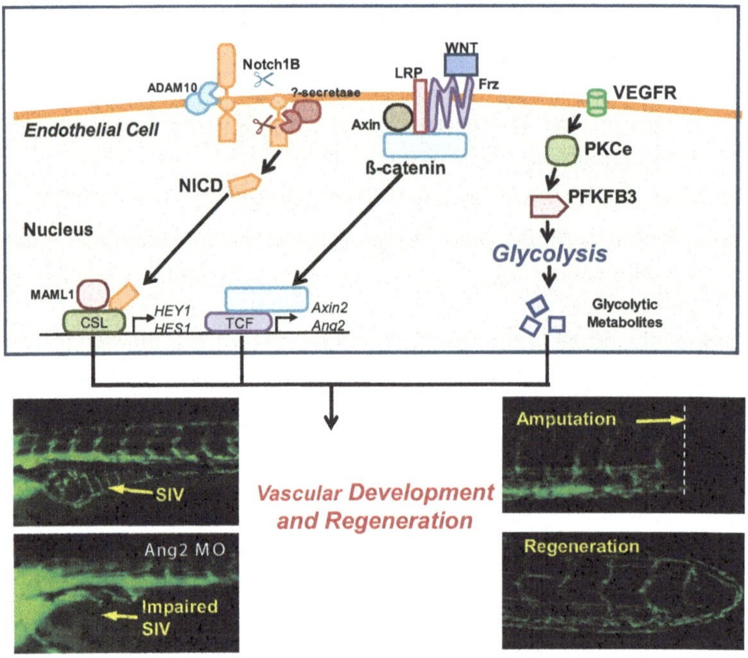

Figure 6.2 Mechanosensitive signal pathways underlying zebrafish vascular development and regeneration. Hemodynamic shear stress activates multiple mechanical and metabolic pathways, including Notch, Wnt/Ang-2, and PKCε/PFKFB3, to modulate vascular development and regeneration in zebrafish. Adapted from Ref. [27b]. Copyright © 2019 Karger Publishers, Basel, Switzerland.

Hemodynamic blood flow exerts shear stress, periodic stretch, and hydrostatic pressure on the endothelium [40, 41]. While cyclic and circumferential stretch is critical for maintaining endothelial function, shear stress is well recognized to mechanically modulate vascular endothelial function via mechanosensitive pathways

[42–44]. For example, Notch signaling is shear sensitive. Unidirectional laminar shear stress induces vascular endothelial growth factor (VEGF)-Notch pathway to drive the expression of the arterial endothelial marker, ephrinB2, while downregulating venous endothelial marker, ephrinB4, in embryonic stem cell driven-murine ECs [45]. Use of embryonic zebrafish allowed for genetic manipulations of blood viscosity to alter the level of endothelial wall shear stress [46] and subsequently Notch signaling in vivo. Our observation with the embryonic zebrafish tail amputation model of vascular injury and repair supported the notion that shear-sensitive Notch signaling plays an important role to facilitate vascular regeneration [47]. The control zebrafish developed vascular regeneration between the dorsal aorta and the dorsal longitudinal anastomotic vessel 3 days after tail amputation, while inhibition of Notch signaling, such as the ADAM10 inhibitor (GI254023X), to prevent proteolytic cleavage of the Notch extracellular domain and the injection of dominant negative (DN)-Notch1b mRNA resulted in impaired vascular regeneration. As a corollary, injection of NICD mRNA to promote Notch signaling–restored vascular regeneration impaired by GI254023X and DN-Notch1b mRNAs [47].

6.3.2 PKCε/PFKFB3 Pathway in Vascular Regeneration

Protein kinase C isoform epsilon (PKCε) is a family of serine and threonine kinase involved in signal transduction, differentiation, and cell proliferation and migration [48–50]. PKCε is abundantly expressed in ECs and is closely associated with a series of pathways involved in angiogenesis and vascular formation. Flow responsive vascular endothelial growth factor receptor (VEGFR)-endothelial nitric oxide synthase pathway drives the expression of PKCε to maintain endothelial homeostasis and lumen formation [51–53]. Shear-mediated nitric oxide (•NO) production [54, 55] further modulates PKCε expression, which, in turn, attenuates mitochondrial reactive oxygen species following ischemia or reperfusion injury [56–60]. PFKFB3 is a rate-limiting enzyme for glycolysis. Endothelial glycolysis is mechanoresponsive [61], and ECs are highly glycolytic [62]. ECs increase the level of glycolytic flux when switching from quiescence to proliferating state, while glycolytic enzymes, including PFKFB3, localize to lamellipodia [63]. As a critical regulator of glycolysis, PFKFB3 involves lamellipodia and filopodia extension for vessel formation. Our recent observation with embryonic zebrafish suggested that intravascular fluid shear modulates PKCε/PFKFB3 to promote vascular regeneration after injury. In the transgenic *Tg(flk1:GFP)* zebrafish tail amputation model, the control zebrafish developed robust vascular regeneration 3 days postamputation (dpa) while a reduction in viscosity-mediated shear stress via *gata1a* morpholino oligonucleotide (MO) injection delayed vascular repair from 3 dpa to 5 dpa [64]. Knockdown of cardiac troponin T type T2 (*tnnt2a*) with MO as a means of arresting myocardial contractility to force restriction of the blood flow also resulted in impairment of vascular regeneration. Injection of the PKCε mRNA overrides the effect of *gata1a* MO, supporting the theory that flow-driven vascular regeneration is at least partially mediated by PKCε [64].

6.3.3 The Wnt/Ang-2 Pathway in Vascular Development and Regeneration

Extensive studies have demonstrated the role of Ang-1 and Ang-2 in vascular development. Shear stress–mediated Ang-2 in mature vascular endothelium was recently reported to play a role in tubulogenesis [65] and to confer atheroprotection [66]. While Ang-1 is constitutively released by the perivascular cells, Ang-2 is expressed in ECs released from the Weibel–Palade bodies upon signal cues [67, 68]. Ang-2, like Ang-1, binds to endothelial–specific receptor tyrosine kinase 2 (TIE-2) and acts as a negative regulator of Ang-1/TIE-2 signaling to promote angiogenesis [69]. Earlier studies demonstrated that Ang-2 could be released by a mechanical force, such as the endothelial stretch that occurs during hypertension [70]. Our recent work and work from Dr. Jo's lab demonstrated that Ang-2 is a mechanosensitive gene involved in shear stress–mediated tubule formation and migration of ECs [71, 72]. In the mouse artery occlusion model, femoral artery ligation caused a disturbed flow to stimulate Ang-2 expression and arteriogenesis in mice [73].

Canonical Wnt/β-catenin signaling is a pivotal pathway regulating development, cell proliferation, and migration [74]. We further demonstrated that shear stress–stimulated Ang-2 expression is mediated by the canonical Wnt signaling pathway. While a Wnt agonist, Wnt3a, promoted Ang-2 expression, inhibition of Wnt signaling with Dickkopfs-1 (Dkk-1) or IWR-1 inhibited EC migration and tube formation. In the heat-shock-inducible Dkk-1 transgenic (*Tg(hsp70l: Dkk1-GFP)*) zebrafish embryos, Ang-2b (zebrafish Ang-2) expression was downregulated upon heat shock. Ang-2 MO injection into transgenic *Tg(kdrl: GFP)* zebrafish caused downregulation of Ang-2, resulting in the impaired development of subintestinal vessels (SIV) at 72 hours after fertilization. Inhibition of Wnt signaling also inhibited the SIV development that is rescued with Ang-2 mRNA. In the zebrafish tail amputation model, inhibition of Wnt signaling retarded and reduced the rate of vascular regeneration that is also rescued by Ang-2 mRNA [72]. These findings support the notion that the mechanosensitive Wnt/Ang-2 pathway modulates vascular development and regeneration.

6.4 Hemodynamic Fluid Force Promotes Cardiac Development via Mechanosensitive Notch Signaling in Zebrafish

In addition to the transcription factors involved in cardiogenic differentiation, hemodynamic fluid forces also play an essential role in cardiogenesis [75–81]. Occlusion of flow at either the cardiac inflow or the cardia outflow tract resulted in hearts with an abnormal third chamber, diminished looping, and impaired valve formation [76]. During heart development, the myocardium differentiates into two layers: an outer compact zone and an inner trabeculated zone. The trabeculae form a network of branching outgrowths from the myocardial wall [82]. Both trabeculation and compaction are essential for normal cardiac contractile function. A significant reduction in trabeculation is usually associated with ventricular compact zone deficiencies, whereas hypertrabeculation (noncompaction) is closely associated with left ventricular noncompaction [83], one of the most common cardiomyopathies in the pediatric population after dilated and hypertrophic cardiomyopathies [84].

Cardiac trabeculation is a crucial morphogenetic process by which clusters of ventricular cardiomyocytes extrude and expand into the cardiac jelly to form sheet-like projections to enhance cardiac contractility and intraventricular conduction. Previous studies support the critical role of Notch signaling in heart development, including trabeculation [85, 86]. Mechanical force induces proteolysis as a mechanism of mechanotransduction to activate Notch signaling [87, 88]. Shear stress (23 dyne·cm^{-2} at 1 Hz) activates Notch signaling in both human and zebrafish aortic ECs [89]. Furthermore, Masumura et al. have also demonstrated shear stress induces time-dependent Notch signaling in murine embryonic stem cell–derived vascular ECs [45].

To elucidate the role and mechanism of hemodynamic forces on trabeculation via notch signaling, Lee et al. took advantage of genetic manipulation in zebrafish and lowered hemodynamic shear forces via (i) microinjection of *gata1a* MO at one- to four-cell stage to reduce hematopoiesis and viscosity by 90% [90, 91], (ii) microinjection of *troponin T type 2a* (*tnnt2a*) MO to arrest cardiomyocyte contraction in embryos [92, 93], and (3) genetic mutation of the *weak atrium* ^{m58} mutant to inhibit atrial contraction [82, 94]. Lowered hemodynamic forces are companied with the inhibition of Notch signaling and trabeculation. Utilizing the *Tg(flk-1:mCherry, tp1:gfp)* fish line, shear stress-activated Notch signaling is confirmed to be localized to the endocardium. Endocardial activation of notch signaling is required for trabeculation. The zebrafish *clo* mutant, in which the endothelial line was abolished, developed a small and thin ventricle [95–97] along with reduced expression of cardiac Notch ligands, receptor, and target genes compared with that of the wild type [89]. Notch activation in the endocardium results in the transcription of *EephrinB2*, which in turn upregulates *Nrg1* [98]. As a secreted factor, *Nrg1* signals to the adjacent cardiomyocytes to promote trabeculation. Disruption of *Nrg1* expression after ischemic insult impairs cardiac contractility [99], whereas *Nrg1* preconditioning confers cardiac protection from ischemic injury [100]. In parallel, Notch activity in the endocardium activates *BMP10* expression in the adjacent myocytes to promote proliferation [98]. Unlike mouse development, *Nrg1/ErbB2* signaling contributes to both proliferation and differentiation of cardiomyocytes for trabeculation in zebrafish [101]. The advantages of the zebrafish model have enabled researchers to establish the essential role of mechanosensitive Notch signaling in promoting cardiac trabeculation involving ephrinB2, Nrg1, and ErbB2[98, 101, 102] (Fig. 6.3).

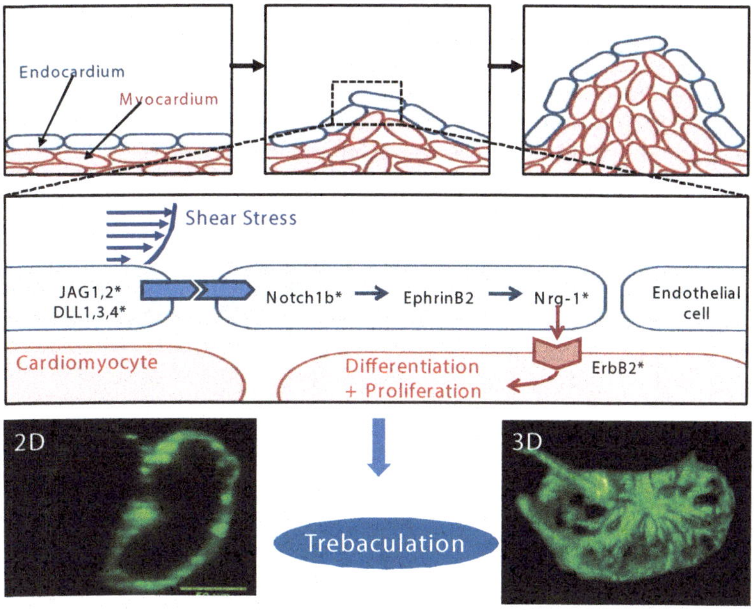

Figure 6.3 Shear stress activation of Notch signaling promotes trabeculation in zebrafish. Shear stress activates Notch signaling in endocardial cells to upregulate the expression of Nrg-1 via ephrinB2. Nrg-1 promotes cardiomyocyte proliferation and differentiation for trabeculation. Adapted from Ref. [27b]. Copyright © 2019 Karger Publishers, Basel, Switzerland.

6.5 Future Perspective

6.5.1 The Regulation of Metabolic Pathways by Mechanical Forces

Metabolomic analyses have led to the discovery of new metabolic biomarkers and therapeutic targets, including polyamines such as spermine for acute stroke; cinnamoylglycine, nicotinamide, and cysteine-glutathione disulfide for renal cancer; and 3-hydroxykynurenine and oxidized glutathione for Parkinson's disease [103–105]. Hemodynamic forces have been shown to modulate mammalian metabolic pathways [106–108] to maintain vascular homeostasis. Unidirectional pulsatile and bidirectional oscillatory flow largely determine the focal but eccentric distribution of vascular oxidative stress [108–110] and post-translational protein modifications. Elucidating flow-mediated metabolomic changes provides an entry point to uncovering metabolites involved in endothelial homeostasis [111, 112], migration [113], vascular development [62], and physical activity [114].

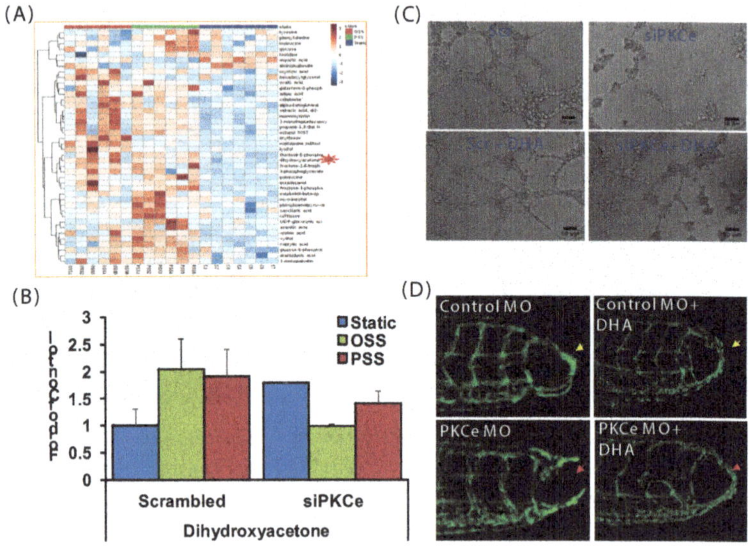

Figure 6.4 Glycolytic metabolite dihydroxyacetone (DHA)-promoted vascular regeneration/repair. (A) Pulsatile shear stress (PSS) and oscillatory shear stress (OSS) modulate a number of metabolites in human aortic endothelial cells (HAEC), including DHA. (B) The level of DHA is mechanosensitive dependent on PKCε. (C) DHA is able to rescue siPKCε-attenuated tube formation in HAEC. (D) PKCε MO–impaired vascular regeneration/repair in zebrafish. Scr/Scrambled: scrambled siRNA; siPKCε: PKCε siRNA; MO: morpholino. Adapted from Ref. [27b]. Copyright © 2019 Karger Publishers, Basel, Switzerland.

Endothelial glycolysis has been shown to be mechanoresponsive as well [61]. Rather than relying on oxidative metabolism for mitochondrial respiration, ECs generate over 80% of their ATP from the glycolytic pathway [115]. A recent study reports that laminar shear stress activates Krüppel-like factor 2 (KLF2) to modulate PFKFB3-mediated glycolysis, mitigating angiogenesis and vessel sprouting [116]. On the other hand, flow-sensitive VEGFR signaling upregulates PFKFB3-driven glycolysis [62, 63]. We have recently demonstrated that shear stress promotes vascular regeneration and repair in zebrafish through VEGFR/PKCε/PFKFB3 signaling, and this effect is at least partially attributed to elevated glycolysis. Shear stress modulates a number of metabolites (Fig. 6.4A), including the elevation of glycolytic metabolite dihydroxyacetone (DHA, $C_3H_6O_3$), which is dependent on flow-sensitive PKCε (Fig. 6.4B). DHA partially rescued tube formation in human aortic ECs (HAECs) and impaired vascular regeneration and repair in zebrafish upon PKCε knockdown (Figs. 6.4C and 6.4D). Hemodynamic forces also modulate SCD1 expression via PPARγ in ECs, a major pathway regulating lipid metabolism [117]. Metabolomic analysis in combination with genetic manipulation of mechanical force in zebrafish would greatly facilitate the investigation of in-depth mechanisms of mechanotransduction in vascular development and pathophysiology.

6.5.2 Interaction and Synergy of Mechanosensitive Pathways

Multiple mechanosensitive pathways such as Notch, Wnt/Ang-2, and PKCε/PFKFB3 signaling have been demonstrated to play important roles in vascular development and regeneration in the zebrafish model. Nakajima et al. recently developed a zebrafish line to model blood flow and monitor the spatiotemporal localization and transcriptional activity of yes-associated protein (YAP) in ECs of living zebrafish. They revealed that blood flow regulates localization and activity of YAP, demonstrating that flow-sensitive endothelial YAP is essential in vessel maintenance [118]. It has recently been reported that flow-sensitive SCD1 generates palmitoleic acid. Palmitoylation of Wnt3a and Wnt5a, the major Wnt signaling agonists, is required for their activation. SCD1 activity also modulates protein palmitoylation of TEAD, the YAP/TAZ coactivator [119]. Furthermore, YAP regulates EC contact-mediated expression of Ang-2 [120]. In cancer cells, SCD1 regulates YAP/TAZ activity by modulating Wnt signaling. Thus mechanosensitive pathways are closely connected and interact with each other. The ability for easy genetic manipulation in zebrafish makes it an excellent model to study the interactions of mechanical force-activated signal pathways in cardiovascular development and homeostasis.

6.5.3 Mechanotransduction of Different Mechanical Forces in Cardiac Morphogenesis

In response to hemodynamic shear forces, myocardial ridges and grooves develop in a wave-like trabecular network in alignment with the direction of the shear stress across the atrioventricular valves [97]. Both trabeculation and compaction are essential for contractile function during cardiac development [82]. We demonstrated that hemodynamic shear forces modulate the initiation of cardiac trabeculation via endocardial Notch-Nrg1 signaling, which activates myocardial *ErbB2* signaling to promote cardiomyocyte proliferation and differentiation to generate contractile force, which also activates Notch signaling [101, 121]. Recently, Han et al. reported that myocardial Notch signaling cell autonomously inhibits *ErbB2* signaling to attenuate trabeculation [122]. The differential effects of hemodynamic shear forces–induced endocardial *Notch-Nrg1* signaling and myocardial contractile force–induced myocardial Notch signaling to inhibit *ErbB2* may modulate the pattern formation of ridges and grooves during cardiac trabeculation. The zebrafish model provides advantages to investigate the pathways activated by different mechanical forces in cardiac pattern formation and morphogenesis.

6.6 Conclusion and Summary

The general strengths of zebrafish for biomedical research are well known. Despite being phylogenetically distant from humans, zebrafish have remarkable similarity in terms of genetics and electrophysiology. In the last two decades, a full array of genetic manipulation techniques have been developed and deployed in zebrafish. The critical mechanosensitive signaling pathways in cardiovascular development and physiology previously studied in mammals have been recapitulated in zebrafish. Studies in the zebrafish model further enhanced our knowledge of the importance of mechanotransduction in cardiovascular development and homeostasis. The application of genomic, transcriptomic, proteomic, and metabolomic analyses in the zebrafish provides a well-characterized platform for the exploration of many biological processes. Furthermore, the zebrafish model allows high-throughput genetic and chemogenomic screening based on critical pathways, making it particularly attractive for therapeutic applications in human diseases.

References

1. Lozano, R., Naghavi, M., Foreman, K., Lim, S., Shibuya, K., Aboyans, V., Abraham, J., Adair, T., Aggarwal, R., Ahn, S. Y., Alvarado, M., Anderson, H. R., Anderson, L. M., Andrews, K. G., Atkinson, C., Baddour, L. M., Barker-Collo, S., Bartels, D. H., Bell, M. L., Benjamin, E. J., Bennett, D., Bhalla, K., Bikbov, B., Bin Abdulhak, A., Birbeck, G., Blyth, F., Bolliger, I., Boufous, S., Bucello, C., Burch, M., Burney, P., Carapetis, J., Chen, H., Chou, D., Chugh, S. S., Coffeng, L. E., Colan, S. D., Colquhoun, S., Colson, K. E., Condon, J., Connor, M. D., Cooper, L. T., Corriere, M., Cortinovis, M., de Vaccaro, K. C., Couser, W., Cowie, B. C., Criqui, M. H., Cross, M., Dabhadkar, K. C., Dahodwala, N., De Leo, D., Degenhardt, L., Delossantos, A., Denenberg, J., Des Jarlais, D. C., Dharmaratne, S. D., Dorsey, E. R., Driscoll, T., Duber, H., Ebel, B., Erwin, P. J., Espindola, P., Ezzati, M., Feigin, V., Flaxman, A. D., Forouzanfar, M. H., Fowkes, F. G., Franklin, R., Fransen, M., Freeman, M. K., Gabriel, S. E., Gakidou, E., Gaspari, F., Gillum, R. F., Gonzalez-Medina, D., Halasa, Y. A., Haring, D., Harrison, J. E., Havmoeller, R., Hay, R. J., Hoen, B., Hotez, P. J., Hoy, D., Jacobsen, K. H., James, S. L., Jasrasaria, R., Jayaraman, S., Johns, N., Karthikeyan, G., Kassebaum, N., Keren, A., Khoo, J. P., Knowlton, L. M., Kobusingye, O., Koranteng, A., Krishnamurthi, R., Lipnick, M., Lipshultz, S. E., Ohno, S. L., Mabweijano, J., MacIntyre, M. F., Mallinger, L., March, L., Marks, G. B., Marks, R., Matsumori, A., Matzopoulos, R., Mayosi, B. M., McAnulty, J. H., McDermott, M. M., McGrath, J., Mensah, G. A., Merriman, T. R., Michaud, C., Miller, M., Miller, T. R., Mock, C., Mocumbi, A. O., Mokdad, A. A., Moran, A., Mulholland, K., Nair, M. N., Naldi, L., Narayan, K. M., Nasseri, K., Norman, P., O'Donnell, M., Omer, S. B., Ortblad, K., Osborne, R., Ozgediz, D., Pahari, B., Pandian, J. D., Rivero, A. P., Padilla, R. P., Perez-Ruiz, F., Perico, N., Phillips, D., Pierce, K., Pope, C. A., 3rd, Porrini, E., Pourmalek, F., Raju, M., Ranganathan, D., Rehm, J. T., Rein, D. B., Remuzzi, G., Rivara, F. P., Roberts, T., De Leon, F. R., Rosenfeld, L. C., Rushton, L., Sacco, R. L., Salomon, J. A., Sampson, U., Sanman, E., Schwebel, D. C., Segui-Gomez, M., Shepard, D. S., Singh, D., Singleton, J., Sliwa, K., Smith, E., Steer, A., Taylor, J. A., Thomas, B., Tleyjeh, I. M., Towbin, J. A., Truelsen, T., Undurraga, E. A., Venketasubramanian, N., Vijayakumar, L., Vos, T., Wagner, G. R., Wang, M., Wang, W., Watt, K., Weinstock, M. A., Weintraub, R., Wilkinson, J. D.,

Woolf, A. D., Wulf, S., Yeh, P. H., Yip, P., Zabetian, A., Zheng, Z. J., Lopez, A. D., Murray, C. J., AlMazroa, M. A. and Memish, Z. A. (2012). Global and regional mortality from 235 causes of death for 20 age groups in 1990 and 2010: a systematic analysis for the Global Burden of Disease Study 2010. *Lancet*, **380**, pp. 2095–2128.

2. Genge, C. E., Lin, E., Lee, L., Sheng, X., Rayani, K., Gunawan, M., Stevens, C. M., Li, A. Y., Talab, S. S., Claydon, T. W., Hove-Madsen, L. and Tibbits, G. F. (2016). The zebrafish heart as a model of mammalian cardiac function. *Rev Physiol Biochem Pharmacol*, **171**, pp. 99–136.
3. MacRae, C. A. and Peterson, R. T. (2015). Zebrafish as tools for drug discovery. *Nat Rev Drug Discov*, **14**, pp. 721–731.
4. Santoriello, C. and Zon, L. I. (2012). Hooked! Modeling human disease in zebrafish. *J Clin Invest*, **122**, pp. 2337–2343.
5. Vornanen, M. and Hassinen, M. (2016). Zebrafish heart as a model for human cardiac electrophysiology. *Channels (Austin)*, **10**, pp. 101–110.
6. Bakkers, J. (2011). Zebrafish as a model to study cardiac development and human cardiac disease. *Cardiovasc Res*, **91**, pp. 279–288.
7. Guyon, J. R., Steffen, L. S., Howell, M. H., Pusack, T. J., Lawrence, C. and Kunkel, L. M. (2007). Modeling human muscle disease in zebrafish. *Biochim Biophys Acta*, **1772**, pp. 205–215.
8. Auer, T. O., Duroure, K., De Cian, A., Concordet, J. P. and Del Bene, F. (2014). Highly efficient CRISPR/Cas9-mediated knock-in in zebrafish by homology-independent DNA repair. *Genome Res*, **24**, pp. 142–153.
9. Hoshijima, K., Jurynec, M. J. and Grunwald, D. J. (2016). Precise editing of the zebrafish genome made simple and efficient. *Dev Cell*, **36**, pp. 654–667.
10. Kimura, Y., Hisano, Y., Kawahara, A. and Higashijima, S. (2014). Efficient generation of knock-in transgenic zebrafish carrying reporter/driver genes by CRISPR/Cas9-mediated genome engineering. *Sci Rep*, **4**, pp. 6545.
11. Nusslein-Volhard, C. and Dahm, R. (2002). *Zebrafish*, Oxford University Press.
12. Poss, K. D., Wilson, L. G. and Keating, M. T. (2002). Heart regeneration in zebrafish. *Science*, **298**, pp. 2188–2190.
13. Raya, Á., Consiglio, A., Kawakami, Y., Rodriguez-Esteban, C. and Izpisúa-Belmonte, J. C. (2004). The zebrafish as a model of heart regeneration. *Cloning Stem Cells*, **6**, pp. 345–351.
14. Kerr, J. P., Robison, P., Shi, G., Bogush, A. I., Kempema, A. M., Hexum, J. K., Becerra, N., Harki, D. A., Martin, S. S. and Raiteri, R. (2015).

Detyrosinated microtubules modulate mechanotransduction in heart and skeletal muscle. *Nat Commun*, **6**, pp. 8526.

15. Rentschler, S., Harris, B. S., Kuznekoff, L., Jain, R., Manderfield, L., Lu, M. M., Morley, G. E., Patel, V. V. and Epstein, J. A. (2011). Notch signaling regulates murine atrioventricular conduction and the formation of accessory pathways. *J Clin Invest*, **121**, pp. 525.

16. Sedmera, D., Reckova, M., deAlmeida, A., Sedmerova, M., Biermann, M., Volejnik, J., Sarre, A., Raddatz, E., McCarthy, R. A. and Gourdie, R. G. (2003). Functional and morphological evidence for a ventricular conduction system in zebrafish and Xenopus hearts. *Am J Physiol Heart Circ Physiol*, **284**, pp. H1152–H1160.

17. Milan, D. J. and MacRae, C. A. (2005). Animal models for arrhythmias. *Cardiovasc Res*, **67**, pp. 426–437.

18. Zhao, Y., Yu, F., Cao, H., Chang, H., Zhang, X., Hsiai, T. K. and Tai, Y.-C. (2013). A wearable percutaneous implant for long term zebrafish epicardial ECG recording. *Solid-State Sensors, Actuators and Microsystems (TRANSDUCERS & EUROSENSORS XXVII), 2013 Transducers & Eurosensors XXVII: The 17th International Conference on*. pp. 756–759.

19. Yu, F., Zhao, Y., Gu, J., Quigley, K. L., Chi, N. C., Tai, Y.-C. and Hsiai, T. K. (2012). Flexible microelectrode arrays to interface epicardial electrical signals with intracardial calcium transients in zebrafish hearts. *Biomed Microdev*, **14**, pp. 357–366.

20. Yu, F., Li, R., Parks, E., Takabe, W. and Hsiai, T. K. (2010). Electrocardiogram signals to assess zebrafih heart regeneration: implication of long QT intervals. *Ann Biomed Eng*, **38**, pp. 2346–2357.

21. Yu, F., Huang, J., Adlerz, K., Jadvar, H., Hamdan, M. H., Chi, N., Chen, J.-N. and Hsiai, T. K. (2010). Evolving cardiac conduction phenotypes in developing zebrafish larvae: implications to drug sensitivity. *Zebrafish*, **7**, pp. 325–331.

22. Cao, H., Yu, F., Zhao, Y., Zhang, X., Tai, J., Lee, J., Darehzereshki, A., Bersohn, M., Lien, C.-L. and Chi, N. C. (2014). Wearable multi-channel microelectrode membranes for elucidating electrophysiological phenotypes of injured myocardium. *Integr Biol*, **6**, pp. 789–795.

23. Sun, P., Zhang, Y., Yu, F., Parks, E., Lyman, A., Wu, Q., Ai, L., Hu, C.-H., Zhou, Q. and Shung, K. (2009). Micro-electrocardiograms to study post-ventricular amputation of zebrafish heart. *Ann Biomed Eng*, **37**, pp. 890–901.

24. Yamamoto, K., Sokabe, T., Watabe, T., Miyazono, K., Yamashita, J. K., Obi, S., Ohura, N., Matsushita, A., Kamiya, A. and Ando, J. (2005). Fluid shear

stress induces differentiation of Flk-1-positive embryonic stem cells into vascular endothelial cells in vitro. *Am J Physiol Heart Circ Physiol*, **288**, pp. H1915– H1924.

25. Huang, H., Nakayama, Y., Qin, K., Yamamoto, K., Ando, J., Yamashita, J. K., Itoh, H., Kanda, K., Yaku, H., Okamoto, Y. N. and Nemoto, Y. (2005). Differentiation from embryonic stem cells to vascular wall cells under in vitro pulsatile flow loading. *J Artif Organs*, **8**, pp. 110–118.
26. Wang, H., Riha, G. M., Yan, S., Li, M., Chai, H., Yang, H., Yao, Q. and Chen, C. (2005). Shear stress induces endothelial differentiation from a murine embryonic mesenchymal progenitor cell line. *Arterioscler Thromb Vasc Biol*, **25**, pp. 1817–1823.
27. (a) Yamamoto, K., Sokabe, T., Watabe, T., Miyazono, K., Yamashita, J. K., Obi, S., Ohura, N., Matsushita, A., Kamiya, A. and Ando, J. (2005). Fluid shear stress induces differentiation of Flk-1-positive embryonic stem cells into vascular endothelial cells in vitro. Am J Physiol Heart Circ Physiol, 288, H1915–H1924; (b) Li, R., Baek, K. I., Chang, C. C., Zhou, B. and Hsiai, T. K. (2019). Mechanosensitive pathways involved in cardiovascular development and homeostasis in zebrafish. *J Vasc Res*, **56**(6), 273–283.
28. Obi, S., Yamamoto, K., Shimizu, N., Kumagaya, S., Masumura, T., Sokabe, T., Asahara, T. and Ando, J. (2009). Fluid shear stress induces arterial differentiation of endothelial progenitor cells. *J Appl Physiol*, **106**, pp. 203–211.
29. Toh, Y.-C. and Voldman, J. (2011). Fluid shear stress primes mouse embryonic stem cells for differentiation in a self-renewing environment via heparan sulfate proteoglycans transduction. *FASEB J*, **25**, pp. 1208–1217.
30. Hellstrom, M., Phng, L. K., Hofmann, J. J., Wallgard, E., Coultas, L., Lindblom, P., Alva, J., Nilsson, A. K., Karlsson, L., Gaiano, N., Yoon, K., Rossant, J., Iruela-Arispe, M. L., Kalen, M., Gerhardt, H. and Betsholtz, C. (2007). Dll4 signalling through Notch1 regulates formation of tip cells during angiogenesis. *Nature*, **445**, pp. 776–780.
31. Lobov, I. B., Renard, R. A., Papadopoulos, N., Gale, N. W., Thurston, G., Yancopoulos, G. D. and Wiegand, S. J. (2007). Delta-like ligand 4 (Dll4) is induced by VEGF as a negative regulator of angiogenic sprouting. *Proc Natl Acad Sci U S A*, **104**, pp. 3219–3224.
32. Bray, S. J. (2016). Notch signalling in context. *Nat Rev Mol Cell Biol*, **17**, pp. 722–735.
33. Krebs, L. T., Xue, Y., Norton, C. R., Shutter, J. R., Maguire, M., Sundberg, J. P., Gallahan, D., Closson, V., Kitajewski, J., Callahan, R., Smith, G. H.,

Stark, K. L. and Gridley, T. (2000). Notch signaling is essential for vascular morphogenesis in mice. *Genes Dev*, **14**, pp. 1343–1352.

34. Joutel, A., Corpechot, C., Ducros, A., Vahedi, K., Chabriat, H., Mouton, P., Alamowitch, S., Domenga, V., Cecillion, M., Marechal, E., Maciazek, J., Vayssiere, C., Cruaud, C., Cabanis, E. A., Ruchoux, M. M., Weissenbach, J., Bach, J. F., Bousser, M. G. and Tournier-Lasserve, E. (1996). Notch3 mutations in CADASIL, a hereditary adult-onset condition causing stroke and dementia. *Nature*, **383**, pp. 707–710.

35. Artavanis-Tsakonas, S., Rand, M. D. and Lake, R. J. (1999). Notch signaling: cell fate control and signal integration in development. *Science*, **284**, pp. 770–776.

36. MacKenzie, F., Duriez, P., Wong, F., Noseda, M. and Karsan, A. (2004). Notch4 inhibits endothelial apoptosis via RBP-Jkappa-dependent and -independent pathways. *J Biol Chem*, **279**, pp. 11657–11663.

37. Quillard, T., Coupel, S., Coulon, F., Fitau, J., Chatelais, M., Cuturi, M. C., Chiffoleau, E. and Charreau, B. (2008). Impaired Notch4 activity elicits endothelial cell activation and apoptosis: implication for transplant arteriosclerosis. *Arterioscler Thromb Vasc Biol*, **28**, pp. 2258–2265.

38. Rostama, B., Peterson, S. M., Vary, C. P. and Liaw, L. (2014). Notch signal integration in the vasculature during remodeling. *Vasc Pharmacol*, **63**, pp. 97–104.

39. Walshe, T. E., Connell, P., Cryan, L., Ferguson, G., Gardiner, T., Morrow, D., Redmond, E. M., O'Brien, C. and Cahill, P. A. (2011). Microvascular retinal endothelial and pericyte cell apoptosis in vitro: role of hedgehog and Notch signaling. *Invest Ophthalmol Visual Sci*, **52**, pp. 4472–4483.

40. Li, Y. S., Haga, J. H. and Chien, S. (2005). Molecular basis of the effects of shear stress on vascular endothelial cells. *J Biomech*, **38**, pp. 1949–1971.

41. Ando, J. and Yamamoto, K. (2011). Effects of shear stress and stretch on endothelial function. *Antioxid Redox Signal*, **15**, pp. 1389–1403.

42. Li, R., Jen, N., Wu, L., Lee, J., Fang, K., Quigley, K., Lee, K., Wang, S., Zhou, B., Vergnes, L., Chen, Y. R., Li, Z., Reue, K., Ann, D. K. and Hsiai, T. K. (2015). Disturbed flow induces autophagy, but impairs autophagic flux to perturb mitochondrial homeostasis. *Antioxid Redox Signal*, **23**, pp. 1207–1219.

43. Lee, J., Packard, R. R. and Hsiai, T. K. (2015). Blood flow modulation of vascular dynamics. *Curr Opin Lipidol*, **26**, pp. 376–383.

44. Cheng, C., Tempel, D., van Haperen, R., de Boer, H. C., Segers, D., Huisman, M., van Zonneveld, A. J., Leenen, P. J., van der Steen, A., Serruys, P. W.,

de Crom, R. and Krams, R. (2007). Shear stress-induced changes in atherosclerotic plaque composition are modulated by chemokines. *J Clin Invest*, **117**, pp. 616–626.

45. Masumura, T., Yamamoto, K., Shimizu, N., Obi, S. and Ando, J. (2009). Shear stress increases expression of the arterial endothelial marker ephrinB2 in murine ES cells via the VEGF-Notch signaling pathways. *Arterioscler Thromb Vasc Biol*, **29**, pp. 2125–2131.

46. Hove, J. R., Koster, R. W., Forouhar, A. S., Acevedo-Bolton, G., Fraser, S. E. and Gharib, M. (2003). Intracardiac fluid forces are an essential epigenetic factor for embryonic cardiogenesis. *Nature*, **421**, pp. 172–177.

47. Baek, K. I., Packard, R. R. S., Hsu, J. J., Saffari, A., Ma, Z., Luu, A. P., Pietersen, A., Yen, H., Ren, B., Ding, Y., Sioutas, C., Li, R. and Hsiai, T. K. (2017). Ultrafine particle exposure reveals the importance of FOXO1/notch activation complex for vascular regeneration. *Antioxid Redox Signal*, **28**(13), pp. 1209-1223.

48. Balciunaite, E., Jones, S., Toker, A. and Kazlauskas, A. (2000). PDGF initiates two distinct phases of protein kinase C activity that make unequal contributions to the G0 to S transition. *Curr Biol*, **10**, pp. 261–267.

49. Racke, F. K., Wang, D., Zaidi, Z., Kelley, J., Visvader, J., Soh, J. W. and Goldfarb, A. N. (2001). A potential role for protein kinase C-epsilon in regulating megakaryocytic lineage commitment. *J Biol Chem*, **276**, pp. 522–528.

50. Mackay, H. J. and Twelves, C. J. (2007). Targeting the protein kinase C family: are we there yet? *Nat Rev Cancer*, **7**, pp. 554–562.

51. Rask-Madsen, C. and King, G. L. (2008). Differential regulation of VEGF signaling by PKC-alpha and PKC-epsilon in endothelial cells. *Arterioscler Thromb Vasc Biol*, **28**, pp. 919–924.

52. Carracedo, S., Sacher, F., Brandes, G., Braun, U. and Leitges, M. (2014). Redundant role of protein kinase C delta and epsilon during mouse embryonic development. *PLoS One*, **9**, pp. e103686.

53. Koh, W., Sachidanandam, K., Stratman, A. N., Sacharidou, A., Mayo, A. M., Murphy, E. A., Cheresh, D. A. and Davis, G. E. (2009). Formation of endothelial lumens requires a coordinated PKCepsilon-, Src-, Pak- and Raf-kinase-dependent signaling cascade downstream of Cdc42 activation. *J Cell Sci*, **122**, pp. 1812–1822.

54. Wang, S., Iring, A., Strilic, B., Albarran Juarez, J., Kaur, H., Troidl, K., Tonack, S., Burbiel, J. C., Muller, C. E., Fleming, I., Lundberg, J. O.,

Wettschureck, N. and Offermanns, S. (2015). P2Y2 and Gq/G11 control blood pressure by mediating endothelial mechanotransduction. *J Clin Invest*, **125**, pp. 3077–3086.

55. Jones, R., Baker, M. B., Weber, M., Harrison, D. G., Bao, G. and Searles, C. D. (2009). Molecular beacons can assess changes in expression and 3'-polyadenylation of human eNOS mRNA. *Am J Physiol Cell Physiol*, **296**, pp. C498–C504.

56. Inagaki, K., Begley, R., Ikeno, F. and Mochly-Rosen, D. (2005). Cardioprotection by epsilon-protein kinase C activation from ischemia: continuous delivery and antiarrhythmic effect of an epsilon-protein kinase C-activating peptide. *Circulation*, **111**, pp. 44–50.

57. Budas, G., Costa, H. M., Jr., Ferreira, J. C., Teixeira da Silva Ferreira, A., Perales, J., Krieger, J. E., Mochly-Rosen, D. and Schechtman, D. (2012). Identification of epsilonPKC targets during cardiac ischemic injury. *Circ J*, **76**, pp. 1476–1485.

58. Cross, H. R., Murphy, E., Bolli, R., Ping, P. and Steenbergen, C. (2002). Expression of activated PKC epsilon (PKC epsilon) protects the ischemic heart, without attenuating ischemic H(+) production. *J Mol Cell Cardiol*, **34**, pp. 361–367.

59. Edmondson, R. D., Vondriska, T. M., Biederman, K. J., Zhang, J., Jones, R. C., Zheng, Y., Allen, D. L., Xiu, J. X., Cardwell, E. M., Pisano, M. R. and Ping, P. (2002). Protein kinase C epsilon signaling complexes include metabolism- and transcription/translation-related proteins: complimentary separation techniques with LC/MS/MS. *Mol Cell Proteomics*, **1**, pp. 421–433.

60. Mayr, M., Liem, D., Zhang, J., Li, X., Avliyakulov, N. K., Yang, J. I., Young, G., Vondriska, T. M., Ladroue, C., Madhu, B., Griffiths, J. R., Gomes, A., Xu, Q. and Ping, P. (2009). Proteomic and metabolomic analysis of cardioprotection: interplay between protein kinase C epsilon and delta in regulating glucose metabolism of murine hearts. *J Mol Cell Cardiol*, **46**, pp. 268–277.

61. Suarez, J. and Rubio, R. (1991). Regulation of glycolytic flux by coronary flow in guinea pig heart. Role of vascular endothelial cell glycocalyx. *Am J Physiol*, **261**, pp. H1994–H2000.

62. De Bock, K., Georgiadou, M. and Carmeliet, P. (2013). Role of endothelial cell metabolism in vessel sprouting. *Cell Metab*, **18**, pp. 634–647.

63. De Bock, K., Georgiadou, M., Schoors, S., Kuchnio, A., Wong, B. W., Cantelmo, A. R., Quaegebeur, A., Ghesquiere, B., Cauwenberghs, S., Eelen, G., Phng, L. K., Betz, I., Tembuyser, B., Brepoels, K., Welti, J., Geudens, I., Segura, I., Cruys, B., Bifari, F., Decimo, I., Blanco, R., Wyns,

S., Vangindertael, J., Rocha, S., Collins, R. T., Munck, S., Daelemans, D., Imamura, H., Devlieger, R., Rider, M., Van Veldhoven, P. P., Schuit, F., Bartrons, R., Hofkens, J., Fraisl, P., Telang, S., Deberardinis, R. J., Schoonjans, L., Vinckier, S., Chesney, J., Gerhardt, H., Dewerchin, M. and Carmeliet, P. (2013). Role of PFKFB3-driven glycolysis in vessel sprouting. *Cell*, **154**, pp. 651–663.

64. Baek, K. I., Li, R., Jen, N., Choi, H., Kaboodrangi, A., Ping, P., Liem, D., Beebe, T. and Hsiai, T. K. (2018). Flow-responsive vascular endothelial growth factor receptor-protein kinase c isoform epsilon signaling mediates glycolytic metabolites for vascular repair. *Antioxid Redox Signal*, **28**, pp. 31–43.

65. Tressel, S. L., Huang, R. P., Tomsen, N. and Jo, H. (2007). Laminar shear inhibits tubule formation and migration of endothelial cells by an angiopoietin-2–dependent mechanism. *Arterioscer Thromb Vasc Biol*, **27**, pp. 2150–2156.

66. Ahmed, A., Fujisawa, T., Niu, X. L., Ahmad, S., Al-Ani, B., Chudasama, K., Abbas, A., Potluri, R., Bhandari, V., Findley, C. M., Lam, G. K. W., Huang, J., Hewett, P. W., Cudmore, M. and Kontos, C. D. (2009). Angiopoietin-2 confers atheroprotection in apoE−/− mice by inhibiting LDL oxidation via nitric oxide. *Circu Res*, **104**, pp. 1333–1336.

67. Ahmed, A. and Fujisawa, T. (2011). Multiple roles of angiopoietins in atherogenesis. *Curr Opin Lipidol*, **22**, pp. 380–385.

68. Thomas, M. and Augustin, H. G. (2009). The role of the angiopoietins in vascular morphogenesis. *Angiogenesis*, **12**, pp. 125–137.

69. Felcht, M., Luck, R., Schering, A., Seidel, P., Srivastava, K., Hu, J., Bartol, A., Kienast, Y., Vettel, C., Loos, E. K., Kutschera, S., Bartels, S., Appak, S., Besemfelder, E., Terhardt, D., Chavakis, E., Wieland, T., Klein, C., Thomas, M., Uemura, A., Goerdt, S. and Augustin, H. G. (2012). Angiopoietin-2 differentially regulates angiogenesis through TIE2 and integrin signaling. *J Clin Invest*, **122**, pp. 1991.

70. Korff, T., Ernst, E., Nobiling, R., Feldner, A., Reiss, Y., Plate, K. H., Fiedler, U., Augustin, H. G. and Hecker, M. (2012). Angiopoietin-1 mediates inhibition of hypertension-induced release of angiopoietin-2 from endothelial cells. *Cardiovasc Res*, **94**, pp. 510–518.

71. Tressel, S. L., Huang, R. P., Tomsen, N. and Jo, H. (2007). Laminar shear inhibits tubule formation and migration of endothelial cells by an angiopoietin-2 dependent mechanism. *Arterioscler Thromb Vasc Biol*, **27**, pp. 2150–2156.

72. Li, R., Beebe, T., Jen, N., Yu, F., Takabe, W., Harrison, M., Cao, H., Lee, J., Yang, H., Han, P., Wang, K., Shimizu, H., Chen, J., Lien, C. L., Chi, N. C. and

Hsiai, T. K. (2014). Shear stress-activated Wnt-angiopoietin-2 signaling recapitulates vascular repair in zebrafish embryos. *Arterioscler Thromb Vasc Biol*, **34**, pp. 2268–2275.

73. Tressel, S. L., Kim, H., Ni, C. W., Chang, K., Velasquez-Castano, J. C., Taylor, W. R., Yoon, Y. S. and Jo, H. (2008). Angiopoietin-2 stimulates blood flow recovery after femoral artery occlusion by inducing inflammation and arteriogenesis. *Arterioscler Thromb Vasc Biol*, **28**, pp. 1989–1995.

74. Logan, C. Y. and Nusse, R. (2004). The Wnt signaling pathway in development and disease. *Annu Rev Cell Dev Biol*, **20**, pp. 781–810.

75. Chen, H., Shi, S., Acosta, L., Li, W., Lu, J., Bao, S., Chen, Z., Yang, Z., Schneider, M. D., Chien, K. R., Conway, S. J., Yoder, M. C., Haneline, L. S., Franco, D. and Shou, W. (2004). BMP10 is essential for maintaining cardiac growth during murine cardiogenesis. *Development*, **131**, pp. 2219–2231.

76. Hove, J. R., Köster, R. W., Forouhar, A. S., Acevedo-Bolton, G., Fraser, S. E. and Gharib, M. (2003). Intracardiac fluid forces are an essential epigenetic factor for embryonic cardiogenesis. *Nature*, **421**, pp. 172–177.

77. Li, J., Hou, B., Tumova, S., Muraki, K., Bruns, A., Ludlow, M. J., Sedo, A., Hyman, A. J., McKeown, L., Young, R. S., Yuldasheva, N. Y., Majeed, Y., Wilson, L. A., Rode, B., Bailey, M. A., Kim, H. R., Fu, Z., Carter, D. A., Bilton, J., Imrie, H., Ajuh, P., Dear, T. N., Cubbon, R. M., Kearney, M. T., Prasad, K. R., Evans, P. C., Ainscough, J. F. and Beech, D. J. (2014). Piezo1 integration of vascular architecture with physiological force. *Nature*, **515**, pp. 279–282.

78. Banjo, T., Grajcarek, J., Yoshino, D., Osada, H., Miyasaka, K. Y., Kida, Y. S., Ueki, Y., Nagayama, K., Kawakami, K., Matsumoto, T., Sato, M. and Ogura, T. (2013). Haemodynamically dependent valvulogenesis of zebrafish heart is mediated by flow-dependent expression of miR-21. *Nat Commun*, **4**, pp. 1978.

79. Ten Dijke, P., Egorova, A. D., Goumans, M. J., Poelmann, R. E. and Hierck, B. P. (2015). TGF-beta signaling in endothelial-to-mesenchymal transition: the role of shear stress and primary cilia. *Sci Signal*, **5**, pp. pt2.

80. Santhanakrishnan, A. and Miller, L. A. (2011). Fluid dynamics of heart development. *Cell Biochem Biophys*, **61**, pp. 1–22.

81. Lucitti, J. L., Jones, E. A., Huang, C., Chen, J., Fraser, S. E. and Dickinson, M. E. (2007). Vascular remodeling of the mouse yolk sac requires hemodynamic force. *Development*, **134**, pp. 3317–3326.

82. Peshkovsky, C., Totong, R. and Yelon, D. (2011). Dependence of cardiac trabeculation on neuregulin signaling and blood flow in zebrafish. *Dev Dyn*, **240**, pp. 446–456.

83. Zhang, W., Chen, H., Qu, X., Chang, C. P. and Shou, W. (2013). Molecular mechanism of ventricular trabeculation/compaction and the pathogenesis of the left ventricular noncompaction cardiomyopathy (LVNC). *Am J Med Genet Part C*, **163C**, pp. 144–156.

84. Nugent, A. W., Daubeney, P. E., Chondros, P., Carlin, J. B., Cheung, M., Wilkinson, L. C., Davis, A. M., Kahler, S. G., Chow, C. W., Wilkinson, J. L., Weintraub, R. G. and National Australian Childhood Cardiomyopathy, S. (2003). The epidemiology of childhood cardiomyopathy in Australia. *N Engl J Med*, **348**, pp. 1639–1646.

85. D'Amato, G., Luxan, G., del Monte-Nieto, G., Martinez-Poveda, B., Torroja, C., Walter, W., Bochter, M. S., Benedito, R., Cole, S., Martinez, F., Hadjantonakis, A. K., Uemura, A., Jimenez-Borreguero, L. J. and de la Pompa, J. L. (2016). Sequential Notch activation regulates ventricular chamber development. *Nat Cell Biol*, **18**, pp. 7–20.

86. Grego-Bessa, J., Luna-Zurita, L., del Monte, G., Bolos, V., Melgar, P., Arandilla, A., Garratt, A. N., Zang, H., Mukouyama, Y. S., Chen, H., Shou, W., Ballestar, E., Esteller, M., Rojas, A., Perez-Pomares, J. M. and de la Pompa, J. L. (2007). Notch signaling is essential for ventricular chamber development. *Dev Cell*, **12**, pp. 415–429.

87. Shergill, B., Meloty-Kapella, L., Musse, A. A., Weinmaster, G. and Botvinick, E. (2012). Optical tweezers studies on Notch: single-molecule interaction strength is independent of ligand endocytosis. *Dev Cell*, **22**, pp. 1313–1320.

88. Gordon, W. R., Zimmerman, B., He, L., Miles, L. J., Huang, J., Tiyanont, K., McArthur, D. G., Aster, J. C., Perrimon, N., Loparo, J. J. and Blacklow, S. C. (2015). Mechanical allostery: evidence for a force requirement in the proteolytic activation of notch. *Dev Cell*, **33**, pp. 729–736.

89. Lee, J., Fei, P., Packard, R. R., Kang, H., Xu, H., Baek, K. I., Jen, N., Chen, J., Yen, H., Kuo, C. C., Chi, N. C., Ho, C. M., Li, R. and Hsiai, T. K. (2016). 4-Dimensional light-sheet microscopy to elucidate shear stress modulation of cardiac trabeculation. *J Clin Invest*, **126**, pp. 1679–1690.

90. Vermot, J., Forouhar, A. S., Liebling, M., Wu, D., Plummer, D., Gharib, M. and Fraser, S. E. (2009). Reversing blood flows act through klf2a to ensure normal valvulogenesis in the developing heart. *PLoS Biol*, **7**, pp. e1000246.

91. Galloway, J. L., Wingert, R. A., Thisse, C., Thisse, B. and Zon, L. I. (2005). Loss of gata1 but not gata2 converts erythropoiesis to myelopoiesis in zebrafish embryos. *Dev Cell*, **8**, pp. 109–116.

92. Arnaout, R., Ferrer, T., Huisken, J., Spitzer, K., Stainier, D. Y., Tristani-Firouzi, M. and Chi, N. C. (2007). Zebrafish model for human long QT syndrome. *Proc Natl Acad Sci U S A*, **104**, pp. 11316–11321.

93. Chi, N. C., Shaw, R. M., Jungblut, B., Huisken, J., Ferrer, T., Arnaout, R., Scott, I., Beis, D., Xiao, T., Baier, H., Jan, L. Y., Tristani-Firouzi, M. and Stainier, D. Y. (2008). Genetic and physiologic dissection of the vertebrate cardiac conduction system. *PLoS Biol*, **6**, pp. e109.

94. Berdougo, E., Coleman, H., Lee, D. H., Stainier, D. Y. and Yelon, D. (2003). Mutation of weak atrium/atrial myosin heavy chain disrupts atrial function and influences ventricular morphogenesis in zebrafish. *Development*, **130**, pp. 6121–6129.

95. Liao, W., Bisgrove, B. W., Sawyer, H., Hug, B., Bell, B., Peters, K., Grunwald, D. J. and Stainier, D. Y. (1997). The zebrafish gene cloche acts upstream of a flk-1 homologue to regulate endothelial cell differentiation. *Development*, **124**, pp. 381–389.

96. Stainier, D. Y., Weinstein, B. M., Detrich, H. W., 3rd, Zon, L. I. and Fishman, M. C. (1995). Cloche, an early acting zebrafish gene, is required by both the endothelial and hematopoietic lineages. *Development*, **121**, pp. 3141–3150.

97. Lee, J., Moghadam, M. E., Kung, E., Cao, H., Beebe, T., Miller, Y., Roman, B. L., Lien, C. L., Chi, N. C., Marsden, A. L. and Hsiai, T. K. (2013). Moving domain computational fluid dynamics to interface with an embryonic model of cardiac morphogenesis. *PLoS One*, **8**, pp. e72924.

98. High, F. A. and Epstein, J. A. (2008). The multifaceted role of Notch in cardiac development and disease. *Nat Rev Genet*, **9**, pp. 49–61.

99. Hedhli, N., Huang, Q., Kalinowski, A., Palmeri, M., Hu, X., Russell, R. R. and Russell, K. S. (2011). Endothelium-derived neuregulin protects the heart against ischemic injury. *Circulation*, **123**, pp. 2254–2262.

100. Fang, S. J., Wu, X. S., Han, Z. H., Zhang, X. X., Wang, C. M., Li, X. Y., Lu, L. Q. and Zhang, J. L. (2010). Neuregulin-1 preconditioning protects the heart against ischemia/reperfusion injury through a PI3K/Akt-dependent mechanism. *Chin Med J*, **123**, pp. 3597–3604.

101. Liu, J., Bressan, M., Hassel, D., Huisken, J., Staudt, D., Kikuchi, K., Poss, K. D., Mikawa, T. and Stainier, D. Y. (2010). A dual role for ErbB2 signaling in cardiac trabeculation. *Development*, **137**, pp. 3867–3875.

102. Lee, J., Fei, P., Sevag Packard, R. R., Kang, H., Xu, H., Baek, K. I., Jen, N., Chen, J., Yen, H., Kuo, C. C., Chi, N. C., Ho, C. M., Li, R. and Hsiai, T. K. (2016). 4-Dimensional light-sheet microscopy to elucidate shear stress modulation of cardiac trabeculation. *J Clin Invest*, **126**, pp. 3158.

103. Jove, M., Mauri-Capdevila, G., Suarez, I., Cambray, S., Sanahuja, J., Quilez, A., Farre, J., Benabdelhak, I., Pamplona, R., Portero-Otin, M. and Purroy, F. (2015). Metabolomics predicts stroke recurrence after transient ischemic attack. *Neurology*, **84**, 36–45.

104. Lewitt, P. A., Li, J., Lu, M., Beach, T. G., Adler, C. H. and Guo, L. (2013) Arizona Parkinson's disease C. 3-hydroxykynurenine and other Parkinson's disease biomarkers discovered by metabolomic analysis. *Mov Disord*, **28**, 1653–1660.

105. Clyne, M. (2012). Kidney cancer: metabolomics for targeted therapy. *Nat Rev Urol*, **9**, pp. 355.

106. Dunn, J., Qiu, H., Kim, S., Jjingo, D., Hoffman, R., Kim, C. W., Jang, I., Son, D. J., Kim, D., Pan, C., Fan, Y., Jordan, I. K. and Jo, H. (2014). Flow-dependent epigenetic DNA methylation regulates endothelial gene expression and atherosclerosis. *J Clin Invest*, **124**, pp. 3187–3199.

107. Frangos, J. A., McIntire, L. V. and Eskin, S. G. (1988). Shear stress induced stimulation of mammalian cell metabolism. *Biotechnol Bioeng*, **32**, pp. 1053–1060.

108. Hwang, J., Ing, M. H., Salazar, A., Lassegue, B., Griendling, K., Navab, M., Sevanian, A. and Hsiai, T. K. (2003). Pulsatile versus oscillatory shear stress regulates NADPH oxidase subunit expression: implication for native LDL oxidation. *Circ Res*, **93**, pp. 1225–1232.

109. Chiu, J. J., Wang, D. L., Chien, S., Skalak, R. and Usami, S. (1998). Effects of disturbed flow on endothelial cells. *J Biomech Eng*, **120**, pp. 2–8.

110. Morbiducci, U., Kok, A. M., Kwak, B. R., Stone, P. H., Steinman, D. A. and Wentzel, J. J. (2016). Atherosclerosis at arterial bifurcations: evidence for the role of haemodynamics and geometry. *Thromb Haemost*, **115**, pp. 484–492.

111. Kluge, M. A., Fetterman, J. L. and Vita, J. A. (2013). Mitochondria and endothelial function. *Circ Res*, **112**, pp. 1171–1188.

112. Sun, X. and Feinberg, M. W. (2015). Regulation of endothelial cell metabolism: just go with the flow. *Arterioscler Thromb Vasc Biol*, **35**, pp. 13–15.

113. Wang, Y., Zang, Q. S., Liu, Z., Wu, Q., Maass, D., Dulan, G., Shaul, P. W., Melito, L., Frantz, D. E., Kilgore, J. A., Williams, N. S., Terada, L. S. and Nwariaku, F. E. (2011). Regulation of VEGF-induced endothelial cell migration by mitochondrial reactive oxygen species. *Am J Physiol Cell Physiol*, **301**, pp. C695–C704.

114. Neufer, P. D., Bamman, M. M., Muoio, D. M., Bouchard, C., Cooper, D. M., Goodpaster, B. H., Booth, F. W., Kohrt, W. M., Gerszten, R. E., Mattson, M. P., Hepple, R. T., Kraus, W. E., Reid, M. B., Bodine, S. C., Jakicic, J. M., Fleg, J. L., Williams, J. P., Joseph, L., Evans, M., Maruvada, P., Rodgers, M., Roary, M., Boyce, A. T., Drugan, J. K., Koenig, J. I., Ingraham, R. H., Krotoski, D., Garcia-Cazarin, M., McGowan, J. A. and Laughlin, M. R. (2015). Understanding the cellular and molecular mechanisms of physical activity-induced health benefits. *Cell Metab*, **22**, pp. 4–11.

115. Culic, O., Gruwel, M. L. and Schrader, J. (1997). Energy turnover of vascular endothelial cells. *Am J Physiol*, **273**, pp. C205–C213.

116. Doddaballapur, A., Michalik, K. M., Manavski, Y., Lucas, T., Houtkooper, R. H., You, X., Chen, W., Zeiher, A. M., Potente, M., Dimmeler, S. and Boon, R. A. (2015). Laminar shear stress inhibits endothelial cell metabolism via KLF2-mediated repression of PFKFB3. *Arterioscler Thromb Vasc Biol*, **35**, pp. 137–145.

117. Qin, X., Tian, J., Zhang, P., Fan, Y., Chen, L., Guan, Y., Fu, Y., Zhu, Y., Chien, S. and Wang, N. (2007). Laminar shear stress up-regulates the expression of stearoyl-CoA desaturase-1 in vascular endothelial cells. *Cardiovasc Res*, **74**, pp. 506–514.

118. Nakajima, H., Yamamoto, K., Agarwala, S., Terai, K., Fukui, H., Fukuhara, S., Ando, K., Miyazaki, T., Yokota, Y., Schmelzer, E., Belting, H. G., Affolter, M., Lecaudey, V. and Mochizuki, N. (2017). Flow-dependent endothelial YAP regulation contributes to vessel maintenance. *Dev Cell*, **40**, pp. 523–536 e6.

119. Chan, P., Han, X., Zheng, B., DeRan, M., Yu, J., Jarugumilli, G. K., Deng, H., Pan, D., Luo, X. and Wu, X. (2016). Autopalmitoylation of TEAD proteins regulates transcriptional output of the Hippo pathway. *Nat Chem Biol*, **12**, pp. 282–289.

120. Choi, H. J., Zhang, H., Park, H., Choi, K. S., Lee, H. W., Agrawal, V., Kim, Y. M. and Kwon, Y. G. (2015). Yes-associated protein regulates endothelial cell contact-mediated expression of angiopoietin-2. *Nat Commun*, **6**, pp. 6943.

121. Samsa, L. A., Givens, C., Tzima, E., Stainier, D. Y., Qian, L. and Liu, J. (2015). Cardiac contraction activates endocardial Notch signaling to modulate chamber maturation in zebrafish. *Development*, **142**, pp. 4080–4091.

122. Han, P., Bloomekatz, J., Ren, J., Zhang, R., Grinstein, J. D., Zhao, L., Burns, C. G., Burns, C. E., Anderson, R. M. and Chi, N. C. (2016). Coordinating cardiomyocyte interactions to direct ventricular chamber morphogenesis. *Nature*, **534**, pp. 700–704.

Chapter 7

Mechanosensitive MicroRNAs in Health and Disease

Myung-Jin Oh, Tzu-Pin Shentu, Daksh Chauhan, and Yun Fang
Department of Medicine, University of Chicago, 5841 S. Maryland Ave., MC 6026, Office M628 Chicago, Illinois 60637
yfang1@medicine.bsd.uchicago.edu

7.1 Introduction

The increasing amount of emerging genetic data have shed light on the functional importance of small noncoding RNAs in mechanotransduction and, ultimately, the progression of a wide range of human diseases [2–4]. Small noncoding RNAs are a subgroup of RNA transcripts that are not translated into proteins. While several noncoding RNAs exist, they are typically classified by size and function (microRNAs [miRNAs], long noncoding RNAs, circular RNAs, Piwi interacting RNAs, enhancer RNAs, etc.) [5, 6]. This chapter is devoted to microRNAs (miRNAs) and their biological roles in cellular mechanotransduction, by which cells exert,

Modern Mechanobiology: Convergence of Biomechanics, Development, and Genomics
Edited by Juhyun Lee, Sharon Gerecht, Hanjoong Jo, and Tzung Hsiai

ISBN 978-981-4800-58-7 (Hardcover), 978-0-429-29483-9 (eBook)
www.jennystanford.com

sense, and convert mechanical stimuli to biochemical responses. Noncoding RNAs indeed contribute to a major fraction of the mammalian transcriptional output, and many of these noncoding RNAs carry out regulatory functions by directing varied stages of gene expression, such as epigenetic modification, messenger RNA (mRNA) stability, and translational control [7, 8]. For instance, miRNAs are highly conserved small RNAs of 19–26 nucleotides that post-transcriptionally inhibit their target genes via mRNA instability and translational suppression. Disruption of the tightly regulated spatial and temporal miRNA expression is implicated in many human diseases, such as cancer, metabolic syndromes, and cardiovascular diseases [9]. Mechanistically, miRNA plays important roles in gene expression, cellular proliferation/differentiation, organ development, tissue homeostasis, and central biological processes in which mechanical cues are instrumental [10]. This chapter summarizes the regulation of miRNAs in vascular cells exposed to two major biomechanical cues, hemodynamic shear stress and cyclic stretch. We also discuss the putative role of miRNAs in regulating the responses of cells to microenvironments.

7.2 MicroRNA in Hemodynamics Sensing

Cells reside in a 3D microenvironment, in which they not only contribute to but also exert and respond to mechanical cues of varying magnitudes, directions, and frequencies [11]. This is especially important in the vasculature, where endothelial cells are constantly exposed to hemodynamics due to the flowing blood right from the time the heart starts beating [12–16]. Large amounts of data have supported the critical role of hemodynamics, particularly the shear forces mainly received by the endothelium, in regulating vascular homeostasis and pathology. Vascular health and vessel remodeling are dynamically regulated by the shear forces acting on the vessels via the endothelial interface and the resulting mechanosensing mechanisms. Meanwhile, the flow-induced mechanotransduction mechanism in the endothelium has been implicated in vascular diseases such as aneurysms, poststenotic dilations, acute lung injury, arteriovenous malformations, aortic valve diseases, and atherosclerosis. The interaction between blood flow and local vessel geometry creates complex spatiotemporal shear stresses on the vessel wall. It has now been well established that endothelial cells are mechanosensitive to different forms of blood flow. There are two types of blood flow patterns that have been established related to atherosclerosis, a disease that classically has been defined as the thickening and hardening of arteries [17]. One of these types of blood flow, which we will term as "disturbed flow" (DF), that typically occurs at vascular sites of atypical geometries such as branches and bifurcations is characterized as complicated patterns of multidirectional hemodynamics at variable frequencies leading to fluid disturbance featuring oscillation, flow reversal, and low time-averaged shear stress (Fig. 7.1). In contrast to DF, blood flow patterns in the straight part of the blood vessels are considered to have a unidirectional flow (UF) and higher time-average shear stress. Both in vitro and in vivo studies have demonstrated that unidirectional laminar flow with high wall shear stress promotes the quiescent endothelial phenotype while DF in the arterial regions of atypical vascular geometry prone to atherosclerosis causatively activates endothelial cells [12–16].

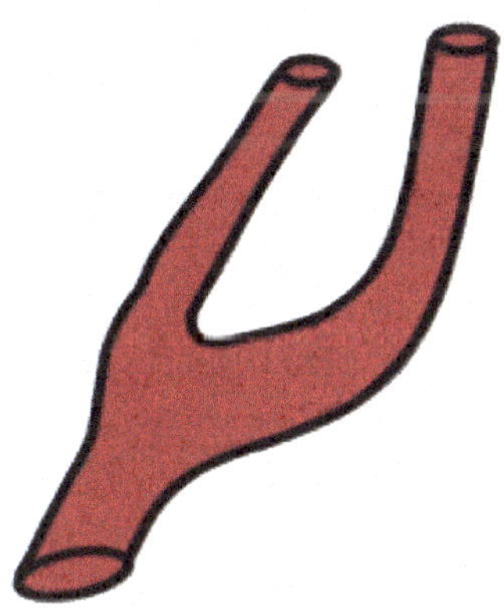

Figure 7.1 Depiction of a bifurcated human carotid artery [1].

While these two types of blood flow are found in the human body, the third and less relevant type of flow pattern used in several studies is one in which endothelial cells are not subjected to any flow pattern, often called a static condition. Studies using microarray data and sequencing results have found multiple miRNAs that can be categorized as proatherogenic or antiatherogenic and are either upregulated or downregulated in UF and DF regions, respectively [18]. While it is generally true that miRNAs that are upregulated in UF regions are downregulated in DF regions and vice versa, a few miRNAs have been shown to have a dual role, depending on the type of shear stress and the type of study involved.

Due to the fact that flow patterns can alter cell signaling in multiple directions on the basis of their flow patterns, we will first discuss the proatherogenic miRNAs that are generally upregulated in DF regions/downregulated in UF regions, then move to miRNAs that are downregulated in DF regions/upregulated in UF regions, and finally cover miRNAs that are modulated by flow but are dependent on the type and experimental conditions, causing some controversy. While the miRNAs highlighted are in no way the complete list, we will try and cover miRNAs that have relevant findings, novel discovery, or clinical outcomes.

- **miR-17-92** cluster: The *MIR17HG* gene encodes a single transcript that folds into six stem loops and comprises six miRNAs—17, 18a, 19a, 19b, 20a, and 92a—based by how close each miRNA was in relation to the others at the same locus. This cluster is also known as "oncomiR-1," given its well-established role in cancer biology and in regulating cell cycle, proliferation, and apoptosis [19]. Later studies demonstrated an important role of the miR-17-92 cluster in vascular homeostasis regulated by biomechanical cues. First, many miRNAs in this cluster, particularly miR-92a, an endothelial cell-enriched miRNA, were shown to be regulated by hemodynamic forces [20, 21]. In vivo and in vitro studies demonstrated that miR-92a expression is significantly upregulated in the endothelium exposed to atherosclerosis-promoting DF and downregulated under UF. Mechanistic investigations showed that miR-92a is a major noncoding RNA that promotes endothelial activation by directly inhibiting transcription factors Krüppel-like factor-2 (KLF2) and KLF4 [20, 21], phospholipid phosphatase 3 (also known as PLPP3 and PPAP2B) [22], Sirtuin 1 [23], and integrin subunits α5 [23, 24].

- **miR-34a**: This particular miRNA is found in atherosclerotic plaques and is critical for endothelial cell senescence [25, 26]. Fan et al. demonstrated that this miRNA was upregulated in oscillatory shear stress and downregulated by UF with high time-averaged shear stress [27]. They furthermore showed that miR-34a alters endothelial cell inflammation by increasing vascular cell adhesion molecule-1 (VCAM-1). VCAM-1 is an important inflammatory molecule that is highly upregulated in atherosclerotic plaques [28]. This, therefore, becomes clinically relevant in miR-34a, potentially altering plaque formation. However, this has currently not been studied. Other studies have also linked miR-34a proliferation in retinal pigment epithelium [29].
- **miR-663**: Microarray studies looking at miRNAs on endothelial cells in DF flow regions found miR-663 to be one of the most upregulated miRNAs [30]. While in endothelial cells there have been links to inflammation [30], miR-663 has been found to be important in vascular smooth muscle cells (SMCs) as well [31]. Interestingly, miR-663 when overexpressed in SMCs caused decreased neointimal formation, an important cause of atherogenesis [31]. Furthermore, miR-663 has been tied to cell proliferation and tumor growth in nasopharyngeal carcinoma cells and pancreatic cancer [32, 33]. This suggests that miR-663 can have multiple roles in clinical outcomes of disease beyond that of mechanosensitive regulation.

Next, we will discuss the antiatherogenic miRNAs that are upregulated in UF regions and downregulated in DF regions.

- **miR-19a** and **miR-23b**: While miR19a is part of the miR-17-92 cluster, endothelial miR-19a appears to have opposing effects in terms of upregulation by UF compared to downregulation like the rest of the miR-17-92 cluster [13, 34]. These studies tested human umbilical vein endothelial cells (HUVECs) compared to HUVECs under static conditions (cells not treated to any flow) [35]. Another miRNA that was highly upregulated in studies testing laminar shear was miR23b [36]. In vascular endothelium, miR-19a was shown to promote cell cycle arrest at G1/S transition by inhibiting cyclinD1 [35] while miR-23b mediates UF-induced G1/Go arrest by suppressing transcription factor E2F1 and Rb hypophosphorylation, consistent with the reported roles of miR-19a and miR-23b in cancer and cell proliferation [36].

- **miR-10a***:* Several reports have found miR-10a to be decreased in DF regions and increased in UF regions both in vitro and in vivo [18, 37] and linked reduced miR-10a expression to increased endothelial inflammation. First, Fang et al. reported that miR-10a serves as an important negative regulator of nuclear factor kappa B (NF-κB) activation by directly suppressing mitogen-activated kinase kinase kinase 7 (MAP3K7), also known as TAK1, and β-transducin repeat-containing gene (*βTRC*) [18], two key regulators of IκBα degradation. Second, Lee et al. showed that endothelial miR-10a reduces VCAM-1 expression via GATA6 suppression [37]. There is growing literature that miR-10a alterations have different roles depending on the cell type [38, 39]. For instance, miR-10a represses proliferation and induces apoptosis in the ovarian granulosa cells [38]; however in acute myeloid leukemia cell studies decreased miR-10a caused increased apoptosis, suggesting miR-10a can cause opposing functions on different cellular pathways [39, 40].

Finally, a look at the dual-modulated miRNAs that have been found increased in both UF and DF depending on the reported conditions. This sets up a controversial set of miRNAs as it is unknown how these particular miRNAs function in terms of disease progression.

- **miR-21**: When compared to static conditions, endothelial miR-21 has been reported to be upregulated by both laminar UF and oscillatory DF in a time-dependent manner [41, 42]. The proatherogeneic role of endothelial miR-21 was suggested by its function of suppressing translation, but not transcription, of peroxisome proliferators-activated receptor-α (PPARα) by 3'-UTR targeting [41]. In contrast, the antiatherogenic property of miR-21 was suggested by experiments detecting decreased apoptosis and activated nitric oxide production in endothelium of miR-21 overexpression [42]. Of more clinical relevance, miR-21 is found to have roles in breast, colon, and hematological cancers [2, 43]. Similar to cancer, abdominal aortic aneurism (AAA) has also been tied to cell proliferation and apoptosis [43]. In AAA, miR-21 was once again upregulated.

- **miR-126**: This is one of the most abundant miRNA clusters expressed in vascular endothelium. Schober et al. reported that DF causatively reduces miR-126-5p expression in vivo, leading to increased Notch1 inhibitor delta-like 1 homolog and promoting atherosclerosis [44]. In contrast, Mondadori dos Santos et al. showed that UF confers the anti-inflammatory endothelial phenotype by increasing the expression of miR-126, which suppresses stromal cell–derived factor-1 SDF-1/CXCL12 and VCAM-1 [45]. Nevertheless, administration of miR-126-5p mimics was shown to significantly lessen high fat–induced atherosclerosis in apolipoprotein E knockout ($ApoE^{-/-}$) mice [44].

7.3 MicroRNA in Extracellular Matrix Regulation

In addition to participating in flow-sensing mechanisms, a cohort of miRNAs have been demonstrated to actively regulate the extracellular matrix (ECM) key to mechanobiology. The composition and organization of the ECM are critical in the numerous biological events, such as tissue repair, organ fibrosis, and cellular development [46, 47]. The ECM includes the interstitial connective tissue matrix and the basement membrane with different constitutive proteins. The interstitial matrix is composed of collagen I and fibronectin, which provide an adhesive and structural framework for cell functions [48, 49]. By contrast, the basement membrane is thinner and more compact than the interstitial matrix and mainly consists of collagen IV, laminins, and heparan sulphate proteoglycans [48, 49]. Collagen molecules typically form a triple stranded helix that gives the tensile strength to support the cellular structure. Elastin, which is closely associated with collagen fibers, can provide the recoil balance to tissues that undergo repeated stretch. Formation, deposition, and remodeling of ECM are regulated on multiple levels;

among them, miRNAs have emerged as critical post-transcriptional regulators of ECM gene expression.

- **miR-29** family: The miR-29 family (miR-29a, miR-29b, and miR-29c) is encoded by two separate loci, targeting several ECM mRNA transcripts, including type I and type III collagen [50–52], fibrillin, and elastin [53, 54]. Overexpression of miR-29 ameliorated bleomycin induced fibrosis in mice by inhibiting connective tissue growth factor expression and Smad3 signaling [55]. The capacity of miR-29 to simultaneously target numerous ECM mRNAs suggests miR-29 as a central participant in overseeing collagen-dependent ECM homeostasis in various tissue contexts.
- **let-7**: Similarly, in line with miR-29, let-7d and miR-200 family (miR-200a, 200b, and 200c) exerted antifibrotic effects by targeting transforming growth factor (TGF)-β1-dependent epithelial to mesenchymal transition. Let-7d inhibition caused increases of mesenchymal genes N-cadherin-2, vimentin, and α-smooth muscle actin (ACTA2) in multiple epithelial cell lines. In vivo intratracheal delivery of let-7d antagomir caused the increase of collagen deposition in the lung [56]. Administration of miR-200 mimics, on the other hand, diminished epithelial to mesenchymal transition in the alveolar epithelial cells, resulting in a reduction of ECM accumulation in experimental mouse models [57].
- **miR-21**: Contrary to the role of miR-29, let-7d, and miR-200, ECM synthesis and accumulation were upregulated by miR-21. The MiR-21 precursor induced the transcripts of fibronectin and a-SMA in lung fibrogenesis [58]. This increased expression of miR-21 promoted uncontrolled collagen production by lung fibroblasts. Besides the importance of miR-21 in regulation of ECM deposition, it has been demonstrated in multiple tissue fibrosis models, such as in cardiac, liver, and kidney fibrosis [59–61], showing its unique mechanism in ECM remodeling.
- **miR-17** and **miR-199a-3p**: Fibronectin is a glycoprotein that is essential for wound healing and cellular development [62–64]. Both miR-17 and miR-199a-3p have been shown to target fibronectin. In addition, the expression of miR-17 decreased endothelial cell adhesion, migration, and proliferation in

vitro and led to growth retardation in transgenic mice [65]. Fibronectin and the fibronectin type-III domain containing 3A are two main targets that were repressed by miR-17. MiR-199a-3p primarily targeted 3'-UTR of fibronectin and versican [66]. Versican is an ECM proteoglycan that impacts cardiac development, cell migration, and inflammation [67–69]. Several miRNAs (including miR-138, miR-136, miR-144, miR-133a, and miR-431) were shown to regulate versican expression [66, 70, 71].

In summary, miRNAs regulate ECM functions by targeting pools of ECM proteins and subsequently affect cell activity and phenotype. This fine-tuned regulation further highlighted ECM integrity in controlling cellular function as well as the delicate impact of miRNAs on the governance of ECM activity and mechanosensing mechanisms.

7.4 MicroRNA in Stretch Sensing

The cardiovascular system represents two distinct activated cellular layers, the endothelium and SMCs, which are constantly exposed to the mechanical stress, such as shear stress, cyclic stretch, and strain. SMCs, as well as the endothelium, withstand the mechanical stretch increasing in the systolic phase of the cardiac cycle and reducing during the diastole phase. SMCs, in particular, are mechanically sensitive to the cyclic stretch. Low-magnitude of cyclic stretch (5%–10%) is considered a physiological condition, whereas high-magnitude (18% or above) stretch is pathologically hypertensive [72, 73]. Several miRNAs are indicative of SMCs' mechanical contractility and their contribution to the vascular health [74]. It's shown that the control of the vascular tone requires the miRNA-143 and miRNA-145 cluster [75–77] and their communication from the endothelium to SMCs [78].

- **miRNA-143** and **miRNA-145**: These two miRNAs maintain SMCs' differentiation in two manners, first, by increasing myocardin and the potentiating serum response factor–myocardin complex and second, by repressing myocardin competitor, Elk-1. In addition, miR-145 inhibits KLF4 and calmodulin kinase II-delta, promoting the SMC differentiation independent of SRF-myocardin effects. Circulating miR-

143/miR-145 in microvesicles secreted from endothelium recapitulates their function in SMCs, contributing to the SMC phenotype. As shown in a study by Hergenreider et al. [78], endothelial cells in response to shear stress activated miR-143/miR-145 enrichment in the extracellular vesicles and reduced Elk-1 and KLF4 expression in SMCs. MiR-1, another key factor in regulating SMC contractile phenotype, was reported to repress KLF4 expression [79]. Inhibition of KLF4 expression increased α-smooth muscle actin and smooth muscle myosin, indicating SMC differentiation. Pulmonary arterial smooth muscle cells displaying increased miR-143/miR-145 correlate with myocardin expression in SMC differentiation. TGF β and bone morphine genetic protein 4 (BMP4) are the signaling molecules upstream of miR-143/miR-145 in maintaining the SMC phenotype [80]. Stretch-induced SMC contractile differentiation was significantly impaired in miR-143/miR-145 knockout mice [81].

- **miR-19b, miR-347, miR-520, miR-568,** and **miR-1290**: Excess mechanical stress leads to cellular deformation and accelerates pathological development. Mechanical stress of cyclic stretch (20% stretch, 20 cycles/min.) to lung epithelium induces epithelial cells to transition to a mesenchymal linage (EMT), showcasing a decrease of cytokeratin-8, E-cadherin, and surfactant protein B as well as an increased expression of vimentin, α-smooth muscle actin, and N-cadherin [82]. During the EMT transition, miR-19b is significantly increased when exposed to cyclic stretch. The inhibitory effects of miR-19b was attributed to an enhanced phosphatidylinositiol-3,4,5-trisphosphate 3 phosphatase pathway. Enhanced cyclic stretch in the lung endothelium also signifies an increase of vascular lung permeability and alveolar edema, which contributes to acute respiratory distress syndrome (ARDS) and ventilator induced lung injury [83]. Pre-B-cell colony-enhancing factor transcription and expression were induced by short-term cyclic stretch (18%). This effect was significantly attenuated by transfection with mimics of miR-374a or miR-568 (40%–60% reductions each), demonstrating the potential of miRNAs in ARDS treatment. Besides, the nonmuscle myosin light chain kinase isoform (nmMLCK) was significantly upregulated under excess mechanical stretch, accompanied by the increase of endothelial permeability [84]. In the same regard, miR-374a/b, miR-520c-3p, and miR-1290 were used to target nmMLCK and subsequently reduced the endothelial permeability. Cyclic stretch of alveoli is characteristic of mechanical ventilation and is postulated to be partly responsible for ventilator-induced lung injury and inflammation [85]. MiR466d-5p and miR466F-3p were upregulated under excess stretch conditions. Using an anti-miRNA inhibitor to specifically knock down 466d and 466f expression decreased the cyclic stretch–induced epithelial permeability.

- **miR-26a**: Mechanical stretch–caused airway remodeling is a characteristic feature observed in the airways of patients having severe asthma and chronic obstructive pulmonary disease. Mohamed et al. demonstrated that cyclic stretch–induced miR-26a serves as a hypertrophic gene [86]. The transcription factor CCAAT enhancer–binding protein directly activates miR-26a expression through the transcriptional machinery. In addition, miR-26a directly targets glycogen synthase kinase-3, an antihypertrophic protein, to enhance hypertrophy in human airway SMCs. Characterized mechanical force is essential for SMC differentiation and directly impacts cellular physiopathological conditions. miRNAs involved in the mechanical stretch are the multilayer players in the transcriptional regulation or gene target characterization. Tentatively, mimics or inhibitory miRNAs in mechanical signaling could lead to therapeutic targets in a diverse range of vascular abnormality.

7.5 MicroRNA in Additional Diseases

As we have highlighted several different miRNAs involved in mechanotransduction- and mechanotransduction-relevant diseases, we will briefly highlight miRNAs' role in a few other diseases. A large assortment of disease states, such as cancer, autoimmune diseases, and skin diseases, have been linked with miRNA [2]. Many types of malignant cancers are tied to mechanisms that control cell growth and differentiation, two critical steps in the development of cancer [87]. Many therapies have therefore attempted to alter cell growth to limit cancers, and miRNAs may be an attractive option. Other organs, such as the kidney, have miRNAs that are critical in their proper development [88]. Water absorption is crucial for kidney function and can be controlled by miR-320a, which in turn regulates aquaporins [89]. Other important kidney functions, such as electrolyte balance and blood pressure, are also regulated by miRNAs [90]. We have previously mentioned miR-34a in retinal epithelium, but other miRNAs have functions in diabetes and other cardiovascular diseases that are not mechanosensitive. To conclude, while many of these miRNAs surround cell growth, proliferation, and apoptosis, the multitude of miRNAs being reported for these different tissue subtypes are varied.

7.6 Targeting Dysregulated Mechanosensitive MicroRNAs in Diseases

Cellular mechanosensing mechanisms converting biomechanical cues into biochemical responses control embryogenesis and tissue homeostasis in which miRNAs are instrumental. Since a single miRNA can impact a complex regulatory network and ultimately the physiological process or disease, targeting miRNAs may be a suitable therapeutic approach to treat human diseases in which mechanosensitive miRNAs are dysregulated and cause disease progression. Although a few miRNA-targeting strategies have shown promise in reaching clinical developments, such as miR-34 mimics in treating cancer and miR-122 antagomirs in treating hepatitis [10], developing mechanointerventions that actively target dysregulated mechanosensing mechanisms remains a challenge. For instance, a major hurdle for nucleotide-based therapies is ineffective intracellular delivery to targeted cells and tissues. Another challenge is the potential for oligonucleotide degradation by RNAses and/or nucleases in circulation or in the endocytic compartment of cells. Systemic administration of therapeutic nucleotides frequently resulted in their predominant accumulation in the liver and unfavorable pharmacokinetic parameters because of rapid in vivo degradation and poor cellular uptake of the nucleotides, leading to low bioavailability in target cells and unwanted side effects in nontarget tissues. Recent advances in oligonucleotide chemistry and nanotechnology may overcome these challenges of miRNA-based therapies. First, nucleotides or the RNA backbone can be chemically modified to enhance the binding affinity, stability, and target modulation effects of miRNA antagomirs and mimics. Second, nanomedicine is an attractive strategy to develop innovative nanocarriers that not only increase the stability of miRNA-modifying agents but also drive active targeting of diseased tissues of interest. Indeed, new lipoparticles and polymeric nanoparticles have been formulated and engineered to deliver miRNA inhibitors against mechanosensitive miR-712 [91] and miR-92a [22], respectively. Integration of our knowledge in miRNA mechanobiology at the molecular and cellular levels in combination with advances in nanotechnology and oligochemistry may have the unique potential to revolutionize future medical practice to treat a wide range of human diseases associated with dysregulated mechanosensing mechanisms.

References

1. Gimbrone, M.A., Jr. and Garcia-Cardena, G. (2013). Vascular endothelium, hemodynamics, and the pathobiology of atherosclerosis. *Cardiovasc Pathol*, **22**(1), pp. 9–15.
2. Ardekani, A. M. and Naeini, M. M. (2010). The role of microRNAs in human diseases. *Avicenna J Med Biotechnol*, **2**(4), pp. 161–179.
3. Bartel, D. P. (2004). MicroRNAs: genomics, biogenesis, mechanism, and function. *Cell*, **116**(2), pp. 281–297.
4. Naeini, M. M. and Ardekani, A. M. (2009). Noncoding RNAs and cancer. *Avicenna J Med Biotechnol*, **1**(2), pp. 55–70.
5. Ha, M. and Kim, V. N. (2014). Regulation of microRNA biogenesis. *Nat Rev Mol Cell Biol*, **15**, pp. 509.
6. Cai, Y., et al. (2009). A brief review on the mechanisms of miRNA regulation. *Genomics Proteomics Bioinf*, **7**, pp. 147–154.
7. Quinn, J. J. and Chang, H. Y. (2016). Unique features of long non-coding RNA biogenesis and function. *Nat Rev Genet*, **17**(1), pp. 47–62.
8. Jonas, S. and Izaurralde, E. (2015). Towards a molecular understanding of microRNA-mediated gene silencing. *Nat Rev Genet*, **16**(7), pp. 421–433.

9. Ambros, V. (2004). The functions of animal microRNAs. *Nature*, **431**(7006), pp. 350–355.

10. Rupaimoole, R. and Slack, F. J. (2017). MicroRNA therapeutics: towards a new era for the management of cancer and other diseases. *Nat Rev Drug Discov*, **16**(3), pp. 203–222.

11. Jaalouk, D. E. and Lammerding, J. (2009). Mechanotransduction gone awry. *Nat Rev Mol Cell Biol*, **10**(1), pp. 63–73.

12. Davies, P. F., et al. (2013). The atherosusceptible endothelium: endothelial phenotypes in complex haemodynamic shear stress regions in vivo. *Cardiovasc Res*, **99**(2), pp. 315–327.

13. Kumar, S., et al. (2014). Role of flow-sensitive microRNAs in endothelial dysfunction and atherosclerosis: mechanosensitive athero-miRs. *Arterioscler Thromb Vasc Biol*, **34**(10), pp. 2206–2216.

14. Gimbrone, M. A., Jr. and Garcia-Cardena, G. (2016). Endothelial cell dysfunction and the pathobiology of atherosclerosis. *Circ Res*, **118**(4), pp. 620–636.

15. Zhou, J., Li, Y. S. and Chien, S. (2014). Shear stress-initiated signaling and its regulation of endothelial function. *Arterioscler Thromb Vasc Biol*, **34**(10), pp. 2191–2198.

16. Hahn, C. and Schwartz, M. A. (2009). Mechanotransduction in vascular physiology and atherogenesis. *Nat Rev Mol Cell Biol*, **10**(1), pp. 53–62.

17. Bergheanu, S. C., Bodde, M. C. and Jukema, J. W. (2017). Pathophysiology and treatment of atherosclerosis: current view and future perspective on lipoprotein modification treatment. *Neth Heart J*, **25**(4), pp. 231–242.

18. Fang, Y., et al. (2010). MicroRNA-10a regulation of proinflammatory phenotype in athero-susceptible endothelium in vivo and in vitro. *Proc Natl Acad Sci U S A*, **107**(30), pp. 13450–13455.

19. Mogilyansky, E. and Rigoutsos, I. (2013). The miR-17/92 cluster: a comprehensive update on its genomics, genetics, functions and increasingly important and numerous roles in health and disease. *Cell Death Differ*, **20**(12), pp. 1603–1614.

20. Fang, Y. and Davies, P. F. (2012). Site-specific microRNA-92a regulation of Kruppel-like factors 4 and 2 in atherosusceptible endothelium. *Arterioscler Thromb Vasc Biol*, **32**(4), pp. 979–987.

21. Wu, W., et al. (2011). Flow-dependent regulation of kruppel-like factor 2 is mediated by microRNA-92a. *Circulation*, **124**(5), pp. 633–641.

22. Wu, C., et al. (2015). Mechanosensitive PPAP2B regulates endothelial responses to atherorelevant hemodynamic forces. *Circ Res*, **117**(4), pp. e41–e53.

23. Chen, Z., et al. (2015). Oxidative stress activates endothelial innate immunity via sterol regulatory element binding protein 2 (SREBP2) transactivation of microRNA-92a. *Circulation*, **131**(9), pp. 805–814.

24. Bonauer, A., et al. (2009). MicroRNA-92a controls angiogenesis and functional recovery of ischemic tissues in mice. *Science*, **324**(5935), pp. 1710–1713.

25. Ito, T., Yagi, S. and Yamakuchi, M. (2010). MicroRNA-34a regulation of endothelial senescence. *Biochem Biophys Res Commun*, **398**(4), pp. 735–740.

26. Raitoharju, E., et al. (2011). miR-21, miR-210, miR-34a, and miR-146a/b are up-regulated in human atherosclerotic plaques in the Tampere Vascular Study. *Atherosclerosis*, **219**(1), pp. 211–217.

27. Fan, W., et al. (2015). Shear-sensitive microRNA-34a modulates flow-dependent regulation of endothelial inflammation. *J Cell Sci*, **128**(1), pp. 70–80.

28. Ley, K. and Huo, Y. (2001). VCAM-1 is critical in atherosclerosis. *J Clin Invest*, **107**(10), pp. 1209–1210.

29. Ruiz, M. A. and Chakrabarti, S. (2013). MicroRNAs: the underlying mediators of pathogenetic processes in vascular complications of diabetes. *Can J Diabetes*, **37**(5), pp. 339–344.

30. Ni, C. W., Qiu, H. and Jo, H. (2011). MicroRNA-663 upregulated by oscillatory shear stress plays a role in inflammatory response of endothelial cells. *Am J Physiol Heart Circ Physiol*, **300**(5), pp. H1762–H1769.

31. Li, P., et al. (2013). MicroRNA-663 regulates human vascular smooth muscle cell phenotypic switch and vascular neointimal formation. *Circ Res*, **113**(10), pp. 1117–1127.

32. Yi, C., et al. (2012). MiR-663, a microRNA targeting p21(WAF1/CIP1), promotes the proliferation and tumorigenesis of nasopharyngeal carcinoma. *Oncogene*, **31**(41), pp. 4421–4433.

33. Zang, W., et al. (2015). miR-663 attenuates tumor growth and invasiveness by targeting eEF1A2 in pancreatic cancer. *Mol Cancer*, **14**(1), pp. 37.

34. Robinson, H. C. and Baker, A. H. (2012). How do microRNAs affect vascular smooth muscle cell biology? *Curr Opin Lipidol*, **23**(5), pp. 405–411.

35. Qin, X., et al. (2010). MicroRNA-19a mediates the suppressive effect of laminar flow on cyclin D1 expression in human umbilical vein endothelial cells. *Proc Natl Acad Sci U S A*, **107**(7), pp. 3240–3244.

36. Wang, K.-C., et al. (2010). Role of microRNA-23b in flow-regulation of Rb phosphorylation and endothelial cell growth. *Proc Natl Acad Sci U S A*, **107**(7), pp. 3234–3239.

37. Lee, D.-Y., et al. (2017). MicroRNA-10a is crucial for endothelial response to different flow patterns via interaction of retinoid acid receptors and histone deacetylases. *Proc Natl Acad Sci U S A*, **114**(8), pp. 2072–2077.

38. Jiajie, T., et al. (2017). Conserved miR-10 family represses proliferation and induces apoptosis in ovarian granulosa cells. *Sci Rep*, **7**, pp. 41304.

39. Molloy, T. J., et al. (2016). miR-10a regulates autophagy and cell cycle in normal karyotype acute myeloid leukaemia. *Blood*, **128**(22), pp. 2862.

40. Bryant, A., et al. (2012). miR-10a is aberrantly overexpressed in Nucleophosmin1 mutated acute myeloid leukaemia and its suppression induces cell death. *Mol Cancer*, **11**, pp. 8.

41. Zhou, J., et al. (2011). MicroRNA-21 targets peroxisome proliferators-activated receptor-alpha in an autoregulatory loop to modulate flow-induced endothelial inflammation. *Proc Natl Acad Sci U S A*, **108**(25), pp. 10355–10360.

42. Weber, M., et al. (2010). miR-21 is induced in endothelial cells by shear stress and modulates apoptosis and eNOS activity. *Biochem Biophys Res Commun*, **393**(4), pp. 643–648.

43. Maegdefessel, L., et al. (2012). MicroRNA-21 blocks abdominal aortic aneurysm development and nicotine-augmented expansion. *Sci Transl Med*, **4**(122), pp. 122ra22.

44. Schober, A., et al. (2014). MicroRNA-126-5p promotes endothelial proliferation and limits atherosclerosis by suppressing Dlk1. *Nat Med*, **20**(4), pp. 368–376.

45. Mondadori dos Santos, A., et al. (2015). miR-126 is involved in vascular remodeling under laminar shear stress. *Biomed Res Int*, **2015**, pp. 497280.

46. Frantz, C., Stewart, K. M. and Weaver, V. M. (2010). The extracellular matrix at a glance. *J Cell Sci*, **123**(Pt 24), pp. 4195–4200.

47. Rutnam, Z. J., Wight, T. N. and Yang, B. B. (2013). miRNAs regulate expression and function of extracellular matrix molecules. *Matrix Biol*, **32**(2), pp. 74–85.

48. Bonnans, C., Chou, J. and Werb, Z. (2014). Remodelling the extracellular matrix in development and disease. *Nat Rev Mol Cell Biol*, **15**(12), pp. 786–801.

49. Mouw, J. K., Ou, G. and Weaver, V. M. (2014). Extracellular matrix assembly: a multiscale deconstruction. *Nat Rev Mol Cell Biol*, **15**(12), pp. 771–785.

50. Maurer, B., et al. (2010). MicroRNA-29, a key regulator of collagen expression in systemic sclerosis. *Arthritis Rheum*, **62**(6), pp. 1733–1743.

51. Wang, B., et al. (2012). Suppression of microRNA-29 expression by TGF-beta1 promotes collagen expression and renal fibrosis. *J Am Soc Nephrol*, **23**(2), pp. 252–265.

52. Cushing, L., et al. (2011). miR-29 is a major regulator of genes associated with pulmonary fibrosis. *Am J Respir Cell Mol Biol*, **45**(2), pp. 287–294.

53. van Rooij, E., et al. (2008). Dysregulation of microRNAs after myocardial infarction reveals a role of miR-29 in cardiac fibrosis. *Proc Natl Acad Sci U S A*, **105**(35), pp. 13027–13032.

54. Zhang, P., et al. (2012). Inhibition of microRNA-29 enhances elastin levels in cells haploinsufficient for elastin and in bioengineered vessels--brief report. *Arterioscler Thromb Vasc Biol*, **32**(3), pp. 756–759.

55. Xiao, J., et al. (2012). miR-29 inhibits bleomycin-induced pulmonary fibrosis in mice. *Mol Ther*, **20**(6), pp. 1251–1260.

56. Pandit, K. V., et al. (2010). Inhibition and role of let-7d in idiopathic pulmonary fibrosis. *Am J Respir Crit Care Med*, **182**(2), pp. 220–229.

57. Yang, S., et al. (2012). Participation of miR-200 in pulmonary fibrosis. *Am J Pathol*, **180**(2), pp. 484–493.

58. Liu, G., et al. (2010). miR-21 mediates fibrogenic activation of pulmonary fibroblasts and lung fibrosis. *J Exp Med*, **207**(8), pp. 1589–1597.

59. Chau, B. N., et al. (2012). MicroRNA-21 promotes fibrosis of the kidney by silencing metabolic pathways. *Sci Transl Med*, **4**(121), pp. 121ra18.

60. Thum, T., et al. (2008). MicroRNA-21 contributes to myocardial disease by stimulating MAP kinase signalling in fibroblasts. *Nature*, **456**(7224), pp. 980–984.

61. Zhang, J., et al. (2015). miR-21 inhibition reduces liver fibrosis and prevents tumor development by inducing apoptosis of CD24+ progenitor cells. *Cancer Res*, **75**(9), pp. 1859–1867.

62. Singh, P., Carraher, C. and Schwarzbauer, J. E. (2010). Assembly of fibronectin extracellular matrix. *Annu Rev Cell Dev Biol*, **26**, pp. 397–419.

63. To, W. S. and Midwood, K. S. (2011). Plasma and cellular fibronectin: distinct and independent functions during tissue repair. *Fibrogen Tissue Repair*, **4**(1), pp. 21.

64. Lenselink, E. A. (2015). Role of fibronectin in normal wound healing. *Int Wound J*, **12**(3), pp. 313–316.

65. Shan, S. W., et al. (2009). MicroRNA MiR-17 retards tissue growth and represses fibronectin expression. *Nat Cell Biol*, **11**(8), pp. 1031–1038.

66. Lee, D. Y., et al. (2009). A 3'-untranslated region (3'UTR) induces organ adhesion by regulating miR-199a* functions. *PLoS One*, **4**(2), pp. e4527.

67. Nandadasa, S., Foulcer, S. and Apte, S. S. (2014). The multiple, complex roles of versican and its proteolytic turnover by ADAMTS proteases during embryogenesis. *Matrix Biol*, **35**, pp. 34–41.

68. Wight, T. N., Kang, I. and Merrilees, M. J. (2014). Versican and the control of inflammation. *Matrix Biol*, **35**, pp. 152–161.

69. Wight, T. N., et al. (2014). Versican and the regulation of cell phenotype in disease. *Biochim Biophys Acta*, **1840**(8), pp. 2441–2451.

70. Morton, S. U., et al. (2008). microRNA-138 modulates cardiac patterning during embryonic development. *Proc Natl Acad Sci U S A*, **105**(46), pp. 17830–17835.

71. Fang, L., et al. (2013). Versican 3'-untranslated region (3'-UTR) functions as a ceRNA in inducing the development of hepatocellular carcinoma by regulating miRNA activity. *FASEB J*, **27**(3), pp. 907–919.

72. Anwar, M. A., et al. (2012). The effect of pressure-induced mechanical stretch on vascular wall differential gene expression. *J Vasc Res*, **49**(6), pp. 463–478.

73. Birukov, K. G. (2009). Cyclic stretch, reactive oxygen species, and vascular remodeling. *Antioxid Redox Signal*, **11**(7), pp. 1651–1667.

74. Welten, S. M., et al. (2016). The multifactorial nature of microRNAs in vascular remodelling. *Cardiovasc Res*, **110**(1), pp. 6–22.

75. Cordes, K. R., et al. (2009). miR-145 and miR-143 regulate smooth muscle cell fate and plasticity. *Nature*, **460**(7256), pp. 705–710.

76. Cheng, Y., et al. (2009). MicroRNA-145, a novel smooth muscle cell phenotypic marker and modulator, controls vascular neointimal lesion formation. *Circ Res*, **105**(2), pp. 158–166.

77. Xin, M., et al. (2009). MicroRNAs miR-143 and miR-145 modulate cytoskeletal dynamics and responsiveness of smooth muscle cells to injury. *Genes Dev*, **23**(18), pp. 2166–2178.

78. Hergenreider, E., et al. (2012). Atheroprotective communication between endothelial cells and smooth muscle cells through miRNAs. *Nat Cell Biol*, **14**(3), pp. 249–256.

79. Xie, C., et al. (2011). MicroRNA-1 regulates smooth muscle cell differentiation by repressing Kruppel-like factor 4. *Stem Cells Dev*, **20**(2), pp. 205–210.

80. Davis-Dusenbery, B. N., et al. (2011). down-regulation of Kruppel-like factor-4 (KLF4) by microRNA-143/145 is critical for modulation of vascular smooth muscle cell phenotype by transforming growth factor-beta and bone morphogenetic protein 4. *J Biol Chem*, **286**(32), pp. 28097–28110.

81. Bhattachariya, A., et al. (2015). Spontaneous activity and stretch-induced contractile differentiation are reduced in vascular smooth muscle of miR-143/145 knockout mice. *Acta Physiol (Oxf)*, **215**(3), pp. 133–143.

82. Mao, P., et al. (2017). MicroRNA-19b mediates lung epithelial-mesenchymal transition via phosphatidylinositol-3,4,5-trisphosphate 3-phosphatase in response to mechanical stretch. *Am J Respir Cell Mol Biol*, **56**(1), pp. 11–19.

83. Adyshev, D. M., et al. (2014). Mechanical stress induces pre-B-cell colony-enhancing factor/NAMPT expression via epigenetic regulation by miR-374a and miR-568 in human lung endothelium. *Am J Respir Cell Mol Biol*, **50**(2), pp. 409–418.

84. Adyshev, D. M., et al. (2013). MicroRNA regulation of nonmuscle myosin light chain kinase expression in human lung endothelium. *Am J Respir Cell Mol Biol*, **49**(1), pp. 58–66.

85. Yehya, N., et al. (2012). MicroRNA modulate alveolar epithelial response to cyclic stretch. *BMC Genomics*, **13**, pp. 154.

86. Mohamed, J. S., Lopez, M. A. and Boriek, A. M. (2010). Mechanical stretch up-regulates microRNA-26a and induces human airway smooth muscle hypertrophy by suppressing glycogen synthase kinase-3beta. *J Biol Chem*, **285**(38), pp. 29336–29347.

87. DeBerardinis, R. J., et al. (2008). The biology of cancer: metabolic reprogramming fuels cell growth and proliferation. *Cell Metab*, **7**(1), pp. 11–20.

88. Griffiths-Jones, S., et al. (2006). miRBase: microRNA sequences, targets and gene nomenclature. *Nucleic Acids Res*, **34**(Database issue), pp. D140–D144.

89. Sepramaniam, S., et al. (2010). MicroRNA 320a functions as a novel endogenous modulator of aquaporins 1 and 4 as well as a potential therapeutic target in cerebral ischemia. *J Biol Chem*, **285**(38), pp. 29223–29230.

90. Suzuki, H. and Saruta, T. (2004). An overview of blood pressure regulation associated with the kidney. *Contrib Nephrol*, **143**, pp. 1–15.

91. Kheirolomoom, A., et al. (2015). Multifunctional nanoparticles facilitate molecular targeting and miRNA delivery to inhibit atherosclerosis in ApoE(-/-) mice. *ACS Nano*, **9**(9), pp. 8885–8897.

Chapter 8

Biomechanics in Cardiac Development Using 4D Light-Sheet Imaging

Victoria Messerschmidt and Juhyun Lee
Joint Department of Bioengineering, UT Arlington/UT Southwestern Medical Center, TX 76010, USA
juhyun.lee@uta.edu

8.1 Introduction

8.1.1 Hemodynamic Shear Stress

Fluid shear stress is generated by a fluids' viscosity, which causes a frictional force that acts tangentially on the surface of endothelial cells. To simplify the shear stress, we imagine that the fluid is trapped between two parallel plates separated by a displacement H (Fig. 8.1A). Assuming that the upper plate is moving, and the bottom plate is fixed, a shear force is required to keep the upper plate in motion with velocity $\mathbf{U}$. This case is called the Couette flow, where

Modern Mechanobiology: Convergence of Biomechanics, Development, and Genomics
Edited by Juhyun Lee, Sharon Gerecht, Hanjoong Jo, and Tzung Hsiai

ISBN 978-981-4800-58-7 (Hardcover), 978-0-429-29483-9 (eBook)
www.jennystanford.com

the shear stress (τ) is the slope of tangential velocity in proportional to the dynamic viscosity (μ).

$$\mu\frac{\mathbf{U}}{2H} \approx \mu\frac{du}{dy}|_{y=-H} = \tau$$

With the assumption of constant pressure (P) throughout the fluid domain, the equation of fluid motion, known as Navier–Stokes equation, is simplified as follows:

$$\frac{dP}{dx} = 0, \mu\frac{d^2u}{dy^2} = 0$$

When the upper plate at $y = H$ is moving with velocity $\mathbf{U}(\mathbf{U}_{\text{top}} = \mathbf{U})$, and the lower plate at $y = -H$ is fixed $\mathbf{U}(\mathbf{U}_{\text{bottom}} = 0)$ fluid motion is linearly defined as

$$\text{u} = \frac{\mathbf{U}}{2}\left(\frac{y}{H} + 1\right)$$

In the case of Poiseuille flow, both upper and lower plates are fixed (Fig. 8.1B). The Navier–Stokes equation for 2D blood flow at a constant pressure applied throughout the fluid domain is defined as

$$\frac{dP}{dx} = \text{constant} \neq 0, \quad \mu\frac{d^2u}{dy^2} = \frac{dP}{dx}.$$

The velocity profile is parabolic, meaning the velocity is maximal at the center and zero at the wall or $y = H$ for the nonslip flow.

$$\mathbf{U} = \frac{1}{2\mu}\frac{dP}{dx}\left(y^2 - H^2\right)$$

In the case of 3D Poiseuille flow in the blood vessel, fluid shear stress is directly proportional to the flow rate of blood (Q) and dynamic viscosity (μ) and inversely proportional to the cube of arterial radius (R).

$$\tau = \frac{4\mu Q}{R^3}$$

Therefore, a small change to the diameter significantly influences wall shear stress (WSS).

Fully developed Poiseuille blood flow seldom occurs in the arterial system in the presence of non-Newtonian properties of blood flow, in which the dynamic viscosity (μ) is not constant. The short distance between branches in response to pulsatile flow prevents fully developed flow. For these reasons, disturbed flow, including oscillatory flow, preferentially and geometrically occurs in the lateral wall of a branching point of the blood vessels or between two ridges in cardiac trabeculae during development (Figs. 8.1C and 8.2).

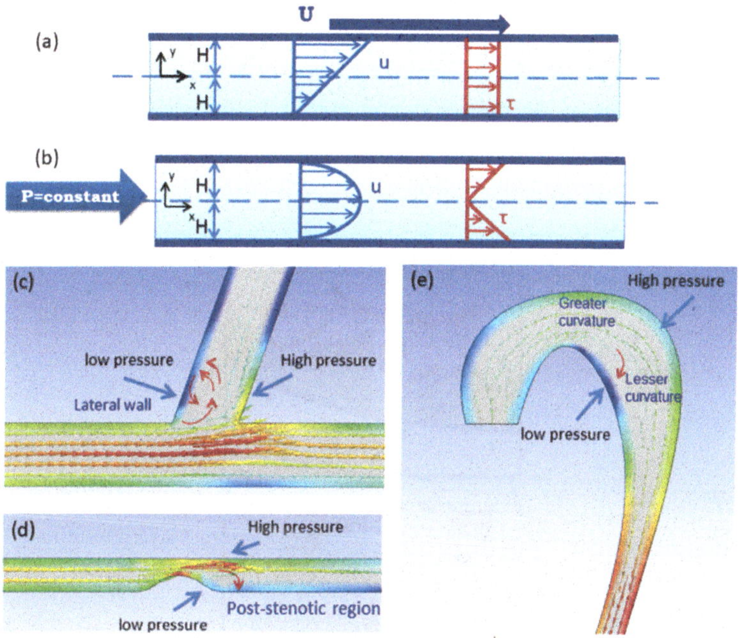

Figure 8.1 Shear stress computational models. (A) Couette flow where the top plate moves at velocity **U** and the bottom plate is fixed. (B) Poiseuille fluid flow model with both top and bottom plates fixed with constant fluid pressure. (C) Fluid vectors modeled at a bifurcation point in a vessel. (D) Fluid vector with modeled stenotic plaque shows a significant change in laminar flow in the post-stenotic region. (E) Fluid flow vectors in a modeled aortic arch. Reproduced from Ref. [62]. Copyright (2015) with permission from Mary Ann Liebert Inc.

This hemodynamic WSS on the endothelial cells is closely related to inflammatory and metabolic effects in the vascular system during the development of atherosclerotic lesions [1]. At the lateral walls of bifurcations, disturbed flow, which includes oscillatory flow (bidirectional and axially misaligned flow), is considered to be an inducer of oxidative stress, which favors the initiation of atherosclerosis. At the medial wall of bifurcation, pulsatile flow (unidirectional and axially aligned flow) downregulates inflammatory cytokines, adhesion molecules, and oxidative stress [2–4]. In addition to the application of cardiac development, for example, blood flow is essential to initiate cardiac trabeculation at distal areas where the blood flows directly onto the ventricular wall opposite of the heart valve [5]. Previous works also demonstrated that the lack of hemodynamic shear stress causes the absence of a valve, bulbus, and disruption of cardiac looping [6]. Blocking both the inlet and the outlet to the heart caused similar phenotypes, which brought about the conclusion that it is the lack of shear forces in the heart, rather than the pressure gradient within the heart that causes these abnormalties [6].

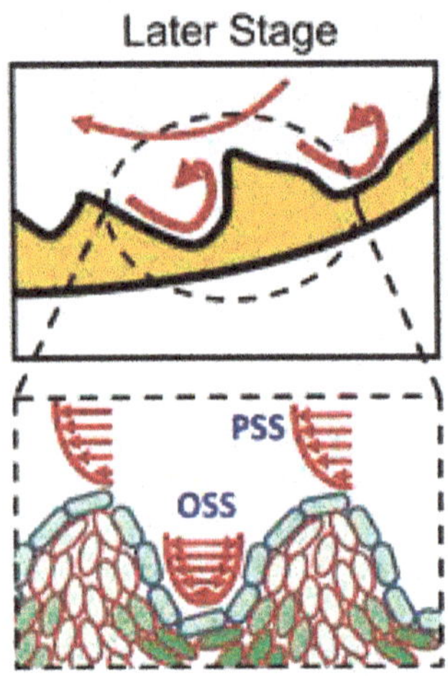

Figure 8.2 Depiction of pulsatile shear stress and oscillatory shear between trabeculae. The top of the trabecular ridges experiences regular pulsatile shear. From the lack of blood flow in the trabecular grooves, low pressure forms, pulling blood down into the groove. The blood then moves down and back up the first ridge and encounters the pulsatile blood flow completing the circular movement that makes oscillatory shear stress. Reproduced from Ref. [59]. Copyright (2012) with permission from the American Society for Clinical Investigation.

8.1.2 Cardiac Trabeculation

Cardiovascular malformation is the leading cause of birth defects in the developed world [7, 8]. In addition, cardiovascular disease is the number one cause of adult morbidity and mortality in the world [8, 9]. In some of the congenital defects, there are irregular trabeculae that cause many different symptoms. In all mammals, trabeculae exist in both the atrium and the ventricle chambers of the heart. As the embryo grows, the heart begins to contract and mature simultaneously with the embryo maturing. Specifically, the trabeculae are highly organized muscular protrusions in the ventricular lumen that form after cardiac looping [7, 8]. Trabeculation occurs in a specific pattern across from the atrioventricular (AV) canal [6, 8, 10]. The protrusions radiate from the stereotypical point across from the AV canal and completely cover the ventricle of zebrafish by 5 days post fertilization (dpf) [7, 10, 11]. However, the trabeculae continue to remodel and mature up to 15 dpf [7, 11]. During this time, the cells along the longitudinal axis of the trabeculae are more differentiated on the luminal side of the trabeculae than on the mural side [8]. The trabeculae stop elongating in the luminal direction and become thicker at their bases to become indistinguishable from the compact myocardium [8]. In fact, the trabeculae become the majority of the myocardial mass instead of the compact layer [8, 10]. As the space between the trabeculae decreases, they form vessels, which become the capillaries [8]. The trabeculae continuously remodel with compact layer development, coronary vasculature, and maturation of the conduction system until they are completely mature [8].

There are currently three theories as to how trabeculae start developing: wall buckling from contractions, active invagination of cardiomyocytes into the lumen, active evagination of endocardium into myocardial layer, or a combination of all three [7, 8, 10]. It has been shown that the Neuregulin and ErbB pathways regulate ventricular trabeculation in zebrafish [7, 10, 11]. The increase in surface area due to the trabeculae allows for more blood oxygenation and nutrition exchange before coronaries develop [7, 11]. Also, the trabeculae regulate cardiac function by increasing cardiac output [7, 8, 11].

During the development and remodeling phase of trabeculation, a series of events that are vulnerable to catastrophic defects occurs. Alterations in cardiogenic events leads to many cardiac diseases like noncompaction cardiomyopathy, diastolic dysfunction, and arrhythmias [8, 12]. Zebrafish are often used in cardiac developmental studies since their gene development and morphogenesis are highly conserved between them and humans [8, 13]. Additionally, zebrafish develop externally, are transparent, and are susceptible to indirect and non-invasive observation of heart development at cellular resolution [8]. It has been shown that hypertrabeculation and hypotrabeculation, an excess and lack of trabeculation, respectively, lead to developmental and functional defects [7, 8, 10, 14, 15]. The reduction in blood flow in *silent heart* (*sih*) or *weak atrium* (*wea*) zebrafish mutants reduces trabeculae, which suggests that proper blood flow is required for maturation of trabeculae [7, 10, 16]. Therefore, the events that occur during trabeculae development are most likely a combination of chemical or genetic signaling and biomechanical force stimulation.

8.1.3 Zebrafish as a Model Animal

Despite having a two-chambered heart and a lack of a pulmonary system, the zebrafish represents an emerging vertebrate model for studying developmental biology (Fig. 8.3A) [20–22]. Its transparency and short organ developmental timeline enable rapid

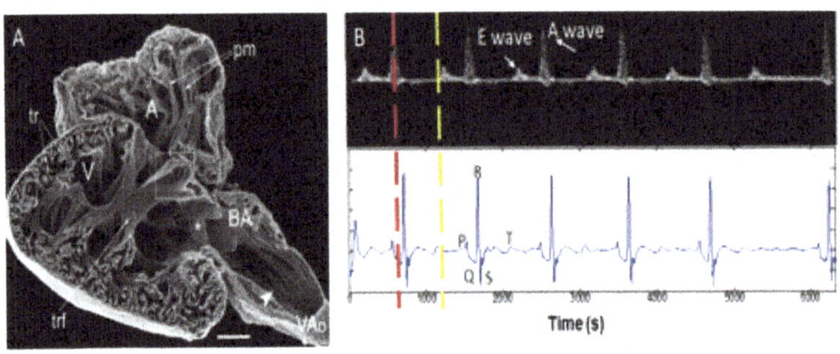

Figure 8.3 Zebrafish cardiac anatomy and physiology. (A) Scanning electron microscopy of the sagittal section of the left half of a 3-month postfertilization zebrafish heart depicting the atrium (A), ventricle (V), bulbus arteriosus (BA), and a portion of the smooth-walled ventral aorta (VAo). The asterisk indicates the bulboventricular valve. The arrowhead identifies one of the elevated ridges along the inner surface of the bulbar wall. tr, trabeculae; trf, trabecular fold; pm, pectinate muscle [17]. (B) Synchronized ECG features co-registered with the hemodynamic events as captured by PW Doppler: P wave in ECG precedes atrial contraction, resulting in A waves in Doppler; T wave in ECG precedes ventricular relaxation, resulting in E waves in Doppler [5, 18, 19].

and high-throughput analysis of developmental stages with optical fluorescent technology [23]. In particular, conservation of a large number of cardiac genes between humans and zebrafish makes it useful in a variety of cardiac research areas [23], such as the feature of cardiac development or regeneration [5, 6, 24–26]. Furthermore, zebrafish have a relatively similar heart rate as humans compared to other popular animal models such as mice or rabbits, and their electrocardiogram (ECG) is nearly the same as humans, exhibiting a P wave for atrial contraction, a QRS complex for ventricular depolarization, and a T wave from repolarization (Fig. 8.3B) [20–22]. Therefore, integrating zebrafish with light-sheet fluorescence microscopy (LSFM) is a powerful research tool of studying the in vivo cardiac developmental mechanism.

8.2 Light-Sheet Technology

8.2.1 Introduction of Light-Sheet Imaging

Imaging of fluorescent samples in biomedical research, in particular a sample's structure in 3D space, has been highlighted and used widely. This technology is specifically used in developmental biology, for instance, in live embryos, to precisely visualize the dynamics of cellular events such as proliferation and migration inside a 3D microenvironment that require 3D analysis for full understanding. Advent of confocal microscope enables the capture high-resolution 3D samples [27]. However, applying this to rapid or dynamic samples, such as the heart, is not applicable due to point-by-point image acquisition technique via pinhole used to block unnecessary light [28]. In addition, multipoint scanning and slow scanning speed affect the sample by inducing photobleaching and photodamage for the sample [28, 29]. Although several other techniques, such as optical projection tomography and optical coherence tomography, have been developed to provide 3D and non-invasive imaging, light-sheet microscopy provides high-resolution fluorescent signals from multiscale samples [30, 31].

More specifically, selective plane illumination microscopy (SPIM) is a subset of light-sheet microscopy that allows high axial resolution and rapid scanning. The SPIM technique uses a simplified fluorescence wide-field microscope set up for detection, which is perpendicular to the illumination axis, or a collimated beam of light [32]. In the illumination axis, cylindrical lenses are used to manipulate the light so that it forms a gaussian beam forcing the beam's waist onto the sample [33].

Compared to other microscopy techniques, SPIM seems to mitigate their setbacks. In confocal microscopy, which is the most commonly used 3D imaging system, the illumination and detection axes overlap in an antiparallel fashion, also called the epifluorescence arrangement [32]. This intersection between detection and illumination causes additional issues due to reflection. Likewise, using one lens results in an anisotropic point spread function (PSF), with the irregularity in the *z* direction [32]. Other imaging modalities that have high resolution are limited by small working distances [34], thus leading to sample size limitations. Techniques allowing for large samples such as computed tomography, positron emission tomography, and magnetic resonance imaging are limited by their spatial resolution and nonspecific contrast [34]. To compensate for these limitations, mechanical slicing of the sample is required [33]. Physically sectioning the sample requires precise cutting and mounting [33], specifically without sample tears, folds, or compressed and stretched areas [34]. Then, 3D reconstruction [34] or image stitching [30] is required, where success relies on the sample thickness [33], sampling rate [34], and computational power. The postprocessing is laborious and very time consuming to obtain a single image. Lastly, because the samples are physically cut, imaging of live and intact samples is not possible [33].

Compared to confocal and wide-field imaging, SPIM has rapid scanning, high axial resolution, low photobleaching and phototoxicity, and spatial localization of cellular events using multiple fluorescent channels [30, 32–35]. Although SPIM does not have perfectly isotropic resolution, the orthogonal orientation of the illumination and detection increases the *z* resolution [30, 32]. The lateral resolution is determined by the detection objective lens, which is similar to regular fluorescence wide-field microscopy [32]. The axial resolution is controlled by the thickness of the light sheet as opposed to the numerical aperture (NA) or a PSF [32]. Even in low NA, the axial resolution in a SPIM setup is much higher than that of a standard confocal microscope simply because of the light-sheet illumination [33]. The low photobleaching and phototoxicity occur in SPIM because only the fluorophores in the thin optical section are exposed and imaged [32]. Conversely, in confocal and conventional microscopy, the entire sample is illuminated, while the camera focuses on a single focal plane, which leads to high exposure [32] (Fig. 8.4). In large samples, SPIM might not be able to detect information that is behind the sample due to light scattering.

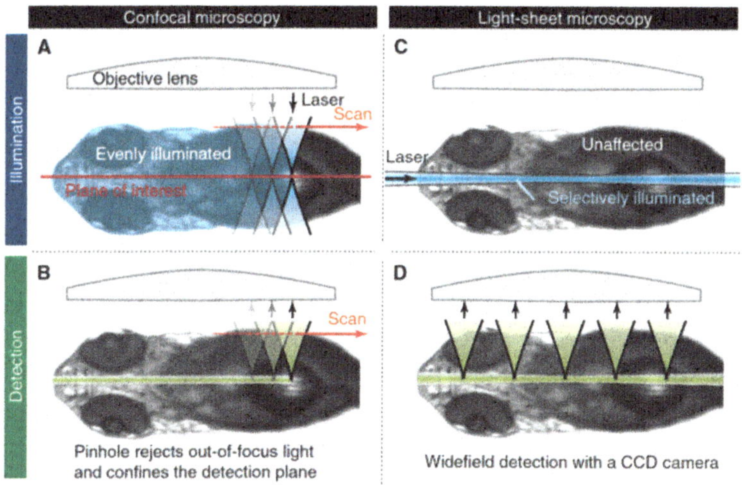

Figure 8.4 Reduction of photobleaching and photodamage in SPIM compared to confocal microscopy. Illustration of the difference in illumination techniques. Confocal microscopy illuminates the entire sample, including the plane of interest (A) and scans across the sample and records the signal (B). Light-sheet microscopy only illuminates the plane of interest (C) and collects all the fluorescent signals at one time (D). Reproduced from Ref. [33]. Copyright (2009) with permission from the Company of Biologists Ltd.

However, the sample can be mounted on a motor, which allows 360° rotation. Images from various angles can be acquired and fused to create a 3D image with a PSF that is the best in all angles [32]. If the sample is large, such as drosophila, fish embryos, or united cell structures [32], the light can be split and directed to illuminate the sample from opposite sides [33]. In confocal imaging, the light signal is lost at a certain depth, where SPIM can compensate with multiple views or multiple directions [33]. In addition, samples do not have to undergo physical sectioning in SPIM, therefore allowing minimally invasive observation overtime with a high signal-to-noise ratio [32, 33]. In fact, the SPIM technique was used to create a timeline of cardiac development in zebrafish [30]. The images were able to pick up slight variations in structure, such as the helical orientation of the cardiomyocytes [30]. One of the biggest advantages of SPIM is that it is easily upgradable [33]. Whenever a faster camera or computer is released, or new algorithms created, small adjustments could incorporate these new technologies [33].

The SPIM technique does require a different procedure for sample preparation—something that not all labs are equipped to handle [33]. Current light-sheet fluorescence microscopy is limited by the pixel size of the camera [36]. However, as mentioned previously, upgrading the camera will eventually occur. Even though fluorescence microscopy has improved penetration, speed, and resolution, there is also a trade-off between the light-sheet thickness and the Rayleigh range for both high spatial resolution and penetration depth [33]. For larger samples, they need to be optically cleared in order to image the entire sample without scattering and absorption [33]. Regardless of these setbacks, the SPIM technique can be further enhanced using postprocessing such as deconvolution, illuminating small volumes of the sample, and integrating subvoxel resolution [33, 36]. The multiple advantages of SPIM are slowly being recognized by various research groups, which will lead to further enhancements of this technology.

8.2.2 Application of Traditional Light-Sheet Imaging

Traditional light-sheet imaging began with neurobiologists and developmental biologists. Brain samples have a high refractive index due to the large concentration of lipids and proteins that naturally occur [37]. To image brain samples, the specimen was placed in a solution with a different refractive index from that of its own [37]. Because of this refractive index mismatch, a large amount of light scattering occurred, causing the brain sample to look opaque [37]. When imaging brains, the neuronal connections that span the entire brain are of interest. Rocha et al. investigated how auditory processing of information could be transferred into motor output [37]. However, cellular resolution of these networks was unattainable at the time. Using both the CUBIC and iDISCO+ tissue-clearing methods, the song bird brains were successfully imaged using fluorescent light-sheet microscopy.

The development of the heart, or other organs, commonly utilizes fluorescent light-sheet microscopy. Chi et al. investigated the development of the cardiac conduction system [16]. Voltage or calcium detection dyes could have been used; however, they do not target specific cells, are limited to live animal experiments, need to be physically delivered to the cells, and cause cellular toxicity reactions [16]. Due to the information obtained using the light-sheet fluorescence microscope, the authors were able to show that the conduction system initially travels linearly across the heart tube. As the heart matures, AV conduction delay and fast conduction develop, giving rise to the His–Purkinje system [16]. Interestingly, after the conduction delay the calcium activation was observed along the trabeculae [16].

Similarly, LSFM was used to model ventricular contraction. Utilizing the Navier–Stokes equations with a moving wall domain, ventricular hemodynamics were found [38]. Previous publications have used particle image velocimetry (PIV) in order to find the biomechanical forces observed by the endothelial cells. However, the analysis fails at the endothelial–blood interface, causing error. The 4D (*xyz* + time) light-sheet images gave high spatiotemporal resolution to successfully model pressure fields and cardiac deformation [38].

Light-sheet microscopy has also been utilized in oceanography. For example, there are spatial interactions between organisms at the micro level that have not successfully been characterized. These structures and interactions are governed by detrital particles, gel-like transparent particles, and aggregated marine snow that is usually disturbed or broken using the current-sampling techniques [39]. A custom light-sheet microscope successfully differentiated between bacterial cells with little blurring. To demonstrate the accuracy, the bacteria concentration was counted using both the light-sheet system and the conventional epifluorescence setup. The epifluorescence determined 12×10^6 bacteria/mL, while the light sheet found 8×10^6 bacteria/mL. The difference was suggested to be from the shadows of the bacteria, the merging of bacteria close together into one pixel, and the lack of detection of bacterial fluorescence due to limited excitation power [39]. Fuchs et al. did find that the light-sheet images were sharper, were simple to use, and had a higher concentration of excitation power compared to other imaging techniques [39].

Light-sheet imaging has also been investigated as a novel diagnosis method. Previously, samples were taken from biopsies that were believed to be diseased. Pathologists would then classify the disease stage on the basis of their own experiences [40, 41]. In clinical applications, microscopy with ultraviolet surface excitation (MUSE) and structured illumination microscopy (SIM) are utilized for their rapid imaging speeds and reduced complexity. However, the techniques are unable to penetrate far into the samples [41]. To test the feasibility of light-sheet microscopy on histology, biopsies were taken from prostate gland or breast tissues suspected of containing malignancies [41]. Fresh prostatectomy specimens could be imaged in less than 1 minute, where a histological analysis could take up to 30 minutes. These samples were first optically cleared and had similar conclusions to the same specimen that had undergone histological analysis [41]. Light-sheet fluorescence microscopy was determined to be as good as histological analysis. However, light-sheet imaging samples allow the observation of the microarchitecture in 3D, so there is no use for slides that would increase cost. Additionally, the same sample could be used later for genetic or molecular tests, and the rapid wide-area imaging is able to image irregular or tilted surfaces [41].

8.2.3 4D Methods to Image in vivo Zebrafish Cardiac Mechanics and Trabeculation

A sequence of images to capture the beating heart was acquired by high-speed scanning via three main steps (Table 8.1): First, the image sequences were captured at a specific section of the heart and were split into multiple periods and processed. Since the zebrafish heart rates were often not in synchrony with the scanning time, we minimized the difference in the least square intensity with respect to the periodic hypothesis (T) by estimating the period of each subsequence [42, 43]. We scanned the fish to ensure that the individual sequence of images cover 4 to 5 cardiac cycles. During data processing, the first and last periods were discarded to ensure data integrity. The images capturing the heart at the same instantaneous moment in different periods were correlated. By using this correlation, we adopted the following cost function to estimate the period:

$$D(z_k,T') = \sum_{m \in Z^2} \sum_{j=1}^{N_\tau - 1} \{ \left| I_m\left[\mathbf{x}_m, z_k, \tau_{i(j)}\right] - I_m\left[x, z_k, \tau_{i(j)}\right] \right|^2 + \left| \tau'_j - \tau'_{j-1} \right|^2 /^{1/2} \tag{8.1}$$

where D denotes the cost function of fitting a period hypothesis, I_m the captured image, τ' the time instance when the image is captured, z_k the z direction index of a slice, and $\mathbf{x}_m$ a vector denoting the pixel index. To increase the accuracy for next steps, we used the image interpolation technique to generate artificial images in between two acquired images [5]. This also enhanced temporal resolution of the 4D process.

Table 8.1 Sequence of 4D reconstruction

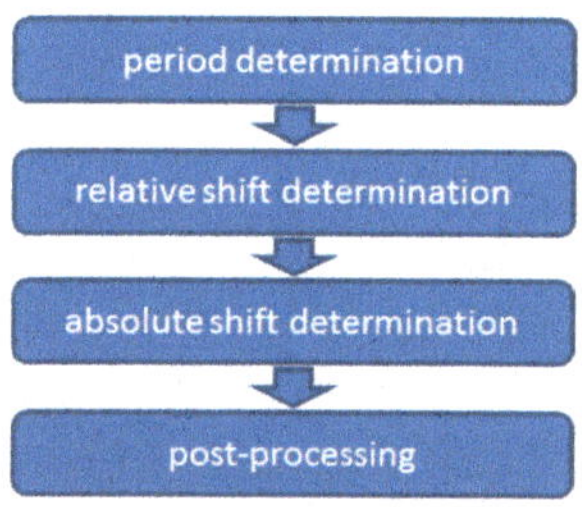

Second, we calculated the relative shift to serve as an indent between image sequences. Despite the identical periods, two image sequences might not be aligned. As mentioned above, the scanning operation proceeded from section to section, and there was an idle time between sections to allow for camera adjustment. When the scanner was ready to capture the next section, the heart was not necessarily in the same cardiac cycle as that of the previous section. The two sections had to be aligned before the 3D structure reconstruction. This phenomenon was called relative shift determination. Two adjacent sections of the heart should have similar appearance. We exploited this similarity and formulated the relative shift determination in the form of quadratic minimization [44]:

$$(s) = \int\int_{R^2} \int_0^L \left| I_m\left(x, z_k, t\right) - I_m\left(x, z_{k'}, t - s\right) \right|^2 dt dx \tag{8.2}$$

where $Q_{k,k'}$ denotes a cost function of fitting a relative shift hypothesis, R the possible spatial neighborhood, L the total time of image capturing, and s the relative shift hypothesis.

Next, we converted the relative shift to the absolute shift with respect to the first image sequence. Although a relative shift could be used to align two adjacent frames, all of the frames were aligned to reconstruct the entire heart. Upon obtaining all relative shifts, a relation to align the next section given the previous section was built. This recurrent relation was transformed into an absolute relation by solving a linear equation (not shown) using use the pseudoinverse approach, as previously described [45].

8.3 Quantification of Hemodynamic Shear Stress

8.3.1 Introduction of CFD

Computational fluid dynamics (CFD) has been used to simulate hemodynamic forces underlying various cardiovascular diseases from unstable plaques to aortic aneurysms [46, 47]. Hemodynamic simulation by CFD provides an understanding of the flow characteristics that could help help clinical decision making, as well as help identify risk factors for potential plaque ruptures. In addition, CFD is widely used to design medical devices such as left ventricular assisted device (LVAD) to evaluate the prototype. Due to variation of severity of the condition among patients, a patient-specific hemodynamic simulation technique is necessary for precise treatment. However, there was still a paucity of information left to understand the molecular events linked with hemodynamic forces simulated by CFD in the cardiac development field. By using computational simulation, the ability to assess detailed hemodynamic information spatially and temporally despite of the limitation of current imaging modalities for a small region is possible.

8.3.2 Combination of Light-Sheet Imaging and CFD

In previous research on cardiovascular hemodynamic simulation, geometry was taken from magnetic resonance imaging (MRI) or computed tomography (CT) scans. In this section, we introduce the moving boundary fluid domain segmentation from optical techniques, such as light-sheet microscopy, for CFD simulation for application in the zebrafish embryos.

The reconstructed 4D zebrafish images are processed through a series of steps to extract the computational model of a beating zebrafish ventricle. A detailed account of the computational modeling framework from zebrafish images to blood flow simulation and WSS computation is presented in Vedula et al. [38]. A brief overview of the framework is discussed here. Firstly, 3D zebrafish images at mid-diastole are segmented using 3D level set segmentation in SimVascular (http://simvascular.github.io/) to create a triangulated surface of the ventricle. Second, to extract the motion of the ventricle, a nonrigid deformable B-spline-based image registration method is employed [48]. In this approach, a source image (e.g., at mid-diastolic phase) is registered to a target image (e.g., at end-diastole) by minimizing a similarity function. A cubic B-spline transformation is used to deform the control points on the source image during registration. A Laplacian-based smoothing operator weighted by a regularization coefficient is added to the similarity function to ensure that the deformations are smooth and non-intersecting. Third, these computed deformations are then used to morph the initial segmented surface to extract the motion of the ventricle. The advantage of using this framework is that the image segmentation is performed only at one instant during the cardiac cycle, whereas the registration is performed directly on images at all other time points to provide the ventricular deformation. This not only saves significant time for model creation (from several hours to a couple of minutes for each registration) but also reduces user intervention by avoiding segmenting 3D images at every time point in the cardiac cycle, followed by a template-based registration to maintain a consistent surface mesh topology. A tetrahedral volume mesh is then created using the TetGen (http://wias-berlin.de/software/tetgen/) open source meshing library that is integrated into SimVascular. Often, the extracted surface was further processed to make a completely closed chamber to avoid fluid leaking during the simulation due to differing intensities of fluorescent signals from each cell or tissue. Therefore, for segmentation-related artifacts such as hole filling, smoothing and extrusion can be done by using a MeshMixer (Autodesk Research Inc.).

Once the computational model has been created, blood flow is simulated by solving Navier–Stokes equations based on arbitrary Lagrangian–Eulerian formulation, assuming blood flow to be incompressible and Newtonian with constant viscosity, using an in-house parallelized finite element solver [49]. The CFD flow solver has been widely used in the cardiovascular research field in multiscale applications [49–51]. The solver is based on stabilized residue-based variational multiscale (RBVMS) finite elements, while the time integration is performed using the generalized alpha method that is stable and second-order accurate. The solver was also employed earlier to simulate hemodynamics in 2D embryonic zebrafish ventricles [52] and was also used in studies of congenital heart disease in humans [53]. Another important caveat for simulating ventricular hemodynamics is the need for dynamic remeshing during the simulation. As the ventricular wall moves and executes large deformations, typical for zebrafish ventricles, the initial mesh degenerates, leading to highly skewed elements with possible self-intersections, and therefore has to be remeshed for the simulation to progress. Typically, remeshing is also followed by an interpolation step to transfer the solution variables from the old mesh to the new mesh. In this case, dynamic remeshing using the TetGen library needs to be employed and the criterion for remeshing is set as the Jacobian of the mesh element to be equal to 0 or negative. The advanced interpolation algorithm helps to reduce the interpolation cost [54].

8.3.3 Application of Zebrafish Cardiac Mechanics and Trabeculation: Morphology

From the CFD simulation, we can determine the area-averaged wall shear stress (AWSS) and time-averaged wall shear stress (TWSS), which are the average shear stress acting on the endocardium over time and on the entire ventricle for one cardiac cycle, respectively (Fig. 8.5A). A *gata1a* morpholino oligomer (MO) injection was performed to reduce viscosity, leading to a reduction in shear stress and concomitant attenuation in trabeculation [5, 26]. As a corollary, *gata1a* MO also reduced the time-averaged AWSS as compared to the control (Fig. 8.5A). Upon simulation with application of the normal viscosity value of blood to the viscosity of *gata1a* MO-

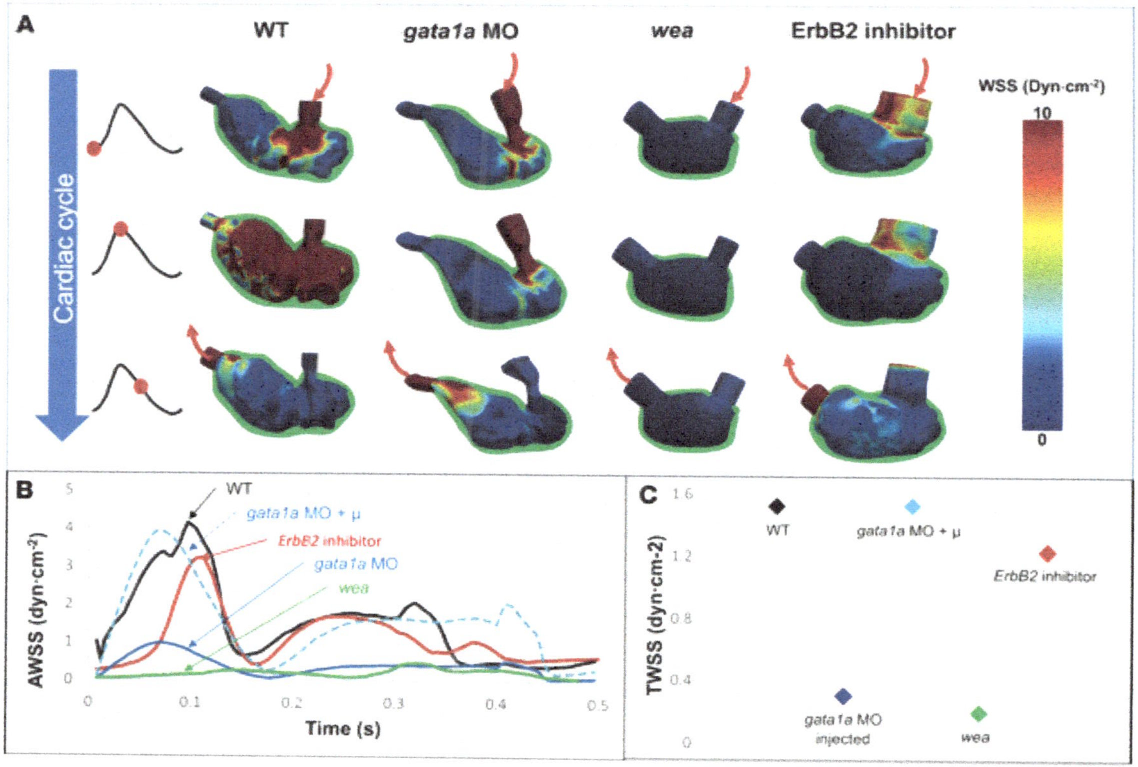

Figure 8.5 Time-dependent 3D computational fluid dynamics simulation of endocardial wall shear stress. (A) Computational fluid dynamic simulations taken from light-sheet images of zebrafish in groups: wild-type (WT), injection with *gata1a* MO, *wea* mutant, and *ErbB2* inhibitor. (B) Average wall shear st⁻ess (AWSS) across the entire cardiac cycle of various experimental groups. (C) Time-averaged wall shear stress (TWSS) of zebrafish groups. Reproduced from Ref. [59]. Copyright (2012) with permission from the American Society for Clinical Investigation.

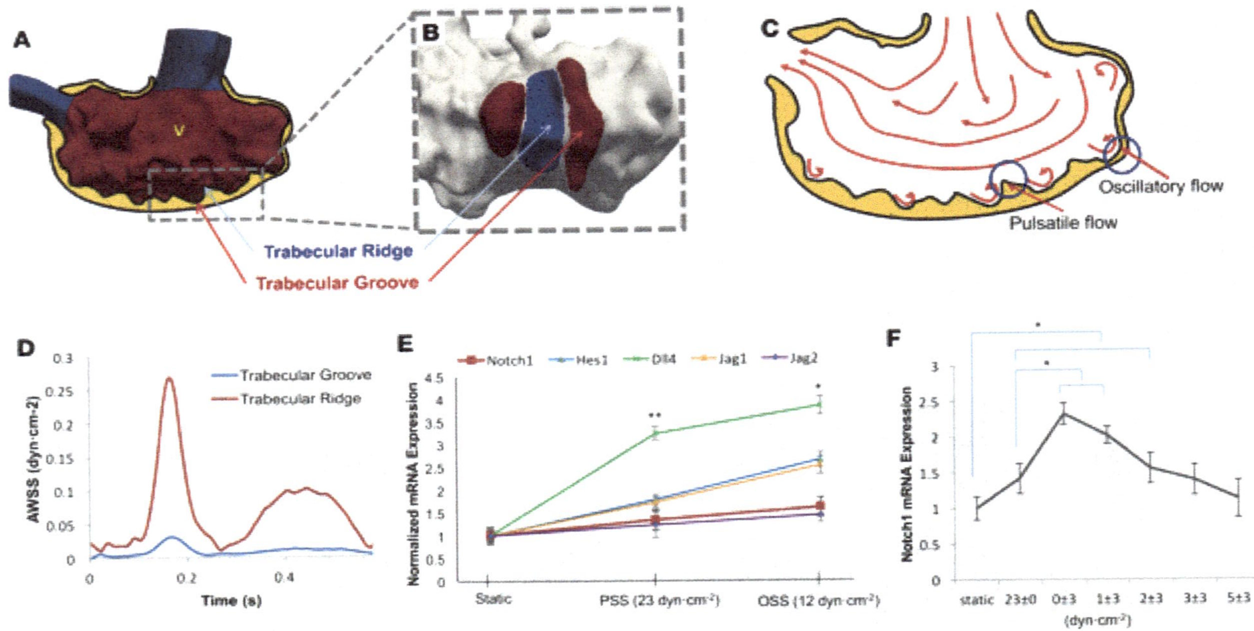

Figure 8.6 Spatiotemporal variation depending on location. (A) Time-dependent 3D computational fluid dynamic simulations of WSS. (B) Different WSS at the trabecular ridges and grooves. (C) Trabecular grooves are subjected to oscillatory shear stress, while trabecular ridges experience pulsatile shear stress. (D) Trabecular ridges measured significantly higher in AWSS. (E) Oscillatory shear stress induced notch-related gene expression more than pulsatile shear stress. (F) Notch mRNA expression at various oscillatory shear stress rates. Reproduced from Ref. [59]. Copyright (2012) with permission from the American Society for Clinical Investigation.

injected ventricular chamber, we observed a near normalization of AWSS to that of the control (Fig. 8.5B). This observation was consistent with the contribution of blood viscosity to shear stress-mediated trabeculation [5]. ErbB signaling inhibition, which directly impacts trabeculation, also attenuated trabeculation, leading to reduced AWSS [10, 11]. While shear stress initiates trabeculation [5], application of cardiac mechanics with CFD simulation further provides a quantitative approach to define the AWSS and TWSS in response to various genetic manipulations for studying cardiac development (Fig. 8.6B,C).

Shear stress acting on the trabecular ridges versus the grooves was further investigated in the wild-type (WT) embryos (Fig. 8.6A–C). At 5 dpf, AWSS acting on the ridges was 3.5-fold higher than that in the grooves. The alternating high and low shear profiles may be implicated in the formation of ridges and grooves in the trabecular network. To assess temporal variation in shear stress modulates trabecular ridge formation, the flow pattern acting on trabecular ridges and grooves was compared. Interestingly, oscillatory shear stress (OSS) was observed in grooves, and pulsatile shear stress (PSS) was acting on trabecular ridges as a result of cardiac contraction. To compare the extent of oscillation of blood flow in the direction and magnitude at a certain area, the oscillatory shear stress index (OSI) was introduced for quantification purposes. The OSI ranged from 0 to 0.5, where 0 indicates a unidirectional net forward flow, and 0.5 reflects an oscillatory flow with no net forward flow. In the WT, flow recirculation developed, as indicated by the elevated OSI profile in the trabecular grooves, and also downstream to the inlet, where flow changes in direction from the atrium to the ventricle forming a vortex (Fig. 8.7). In response to ErbB2 inhibition to attenuate trabeculation, the OSI was mitigated except for the region downstream to the inlet (Fig. 8.7K). In response to *gata 1a* MO injection, there was a reduction in shear stress, attenuation of trabeculation [5], and the OSI profile was reduced. In the *wea* mutants, in which atrial contraction was absent and ventricular trabeculation was attenuated [5], the OSI profile was further reduced. Therefore, trabeculated ventricles promote flow recirculation in the WT. This observation prompted us to assess the implications of kinetic energy dissipation and notch signaling underlying trabeculation and contractile function.

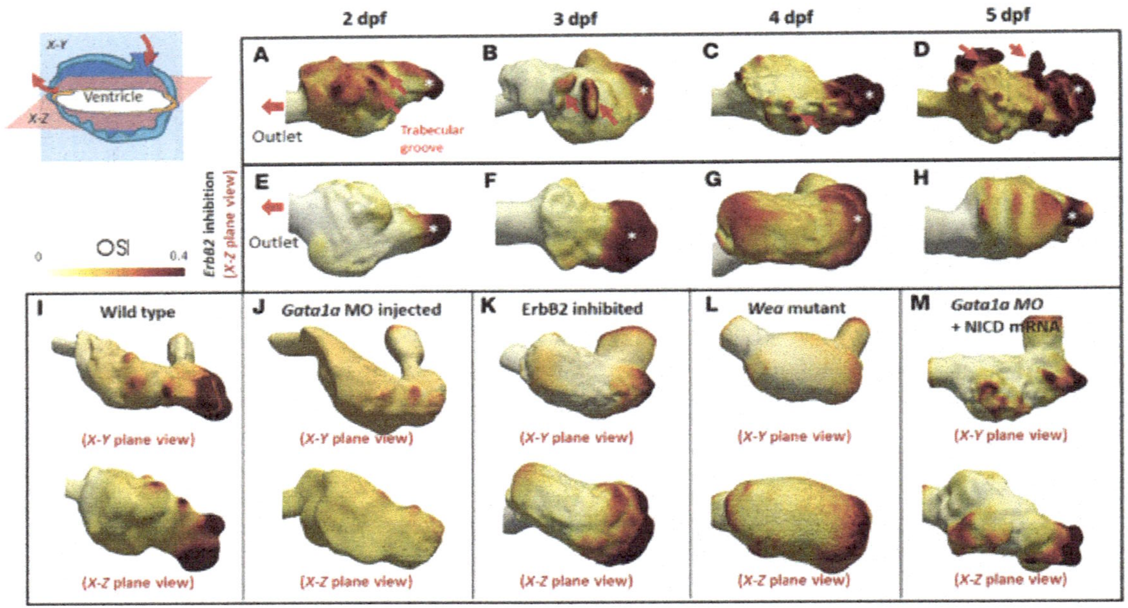

Figure 8.7 Oscillatory shear stress values in various treated or mutant zebrafish. (A and B) 2 and 3 dpf OSI values spread between the ridges and grooves. (C and D) 4 and 5 dpf OSI values became associated with the trabecular network. (E–G) ErbB2 inhibition reduced the trabeculation and OSI at 2, 3, and 4 dpf. (H) At 5 dpf, trabeculae began to form and show OSI values similar to the WT trabecular network. (I) High OSC values are shown at 4 dpf in WT zebrafish. (J) After injection with *gata1a* MO the OSI values significantly decreased. (K) ErbB2-inhibited zebrafish slowed OSS. (L) *wea* mutants had little OSI from lack of atrial contraction. (M) Co-injection of *gata1a* MO and notch intracellular domain (NICD) mRNA partially restored OSI values similar to WT zebrafish. Reproduced from Ref. [59]. Copyright (2012) with permission from the American Society for Clinical Investigation.

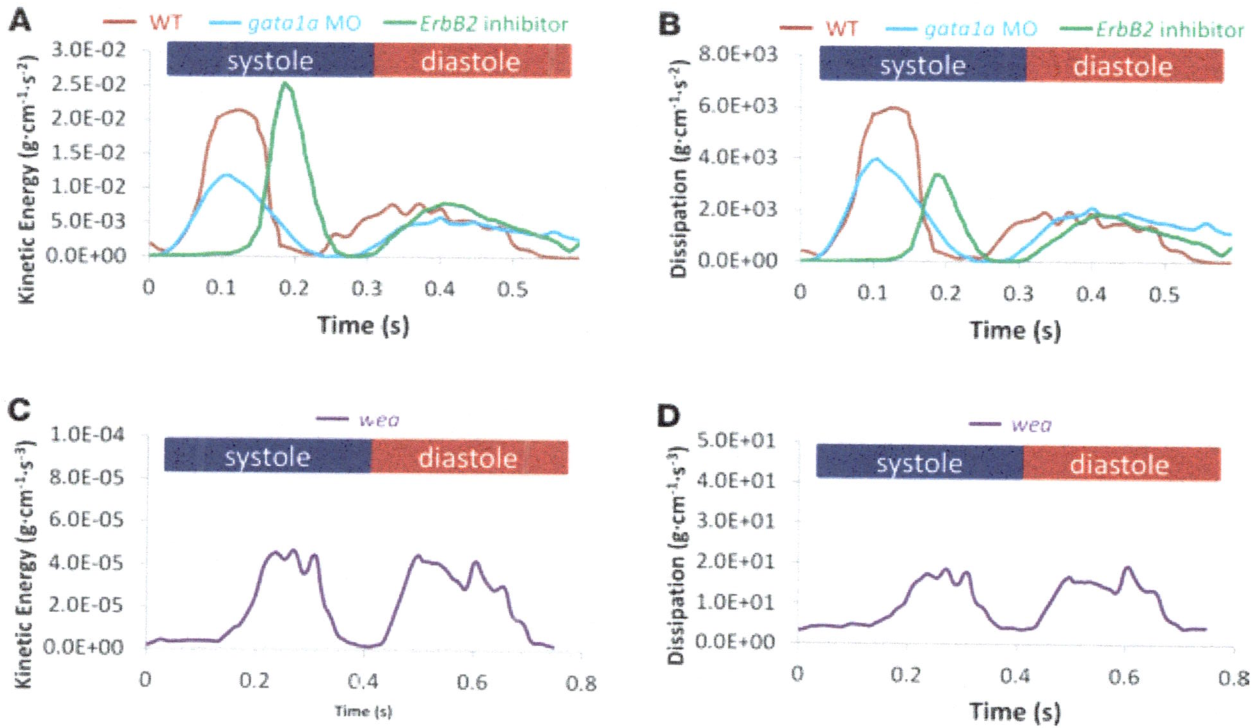

Figure 8.8 Kinetic energy dissipation in trabeculae. (A) The *gata1a* MO had a significantly lower kinetic energy compared to that of WT and ErbB2-inhibited zebrafish. (B) WT zebrafish had much higher energy dissipation than *gata1a* MO and ErbB2-inhibited zebrafish. (C) *wea* mutants had much lower kinetic energy and energy dissipation (D). Reproduced from Ref. [59]. Copyright (2012) with permission from the American Society for Clinical Investigation.

Atrium contraction [18] causes blood to flow across the AV valve, which creates kinetic energy dissipation on the endocardial wall. High kinetic energy dissipation developed in the trabeculated ventricle (WT) as compared to the attenuated trabeculated zebrafish (*gata1a* MO-injected) (Fig. 8.8A). A trabeculated ventricle has higher kinetic energy dissipation developed as compared to the nontrabeculated ventricle (Fig. 8.8B). In addition to the contractile function (Fig. 8.8), this observation suggests that the evolutionary role of trabeculation is to mitigate the endocardial wall from being exposed to high kinetic energies. In the *wea* mutants, lack of atrial contraction resulted in the absence of WSS, leading to nearly zero kinetic energy dissipation to the endocardium. This low kinetic energy dissipation in a nontrabeculated endocardium may be implicated in ventricular remodeling and contractile dysfunction.

8.4 Mechanobiology of Zebrafish Trabeculation

8.4.1 Introduction to Notch Signaling

Notch signaling is essential for trabeculation of the heart. The notch ligand is a membrane-bound protein that transmits a signal to the nucleus once activated. This signaling pathway has been shown to be controlled by a manic fringe-dependent manner. Depending on the concentration of the activated fringe family of glycosyltransferases, the selectivity of notch can change between delta-like (DLL) and jagged (Jag) ligands [55]. The manic fringe enzymes alter the selectivity of the notch receptor by elongating the carbohydrates on the extracellular portion of notch. It also regulates the signals that connect the endocardium to the myocardium for trabeculation and chamber development [55]. Therefore, without notch there may not be a way for the endocardium and myocardium to communicate and initiate trabeculation and normal heart development. When the notch ligand is endocytosed, notch physically changes the active sight so that it is able to bind to ADAM metalloproteases. These allow γ-secretase to cleave the ligand to release the notch intracellular cytoplasmic domain (NICD) [55]. The NICD then travels to the nucleus and enhances the expression of the target gene.

8.4.2 Mechanotransduction of Notch, Including in vitro Cell Studies

Endothelial cells are able to sense the magnitude, direction, amplitude, and frequency of fluid flow in their environment [56]. The mechanical forces applied to the cells affect notch signaling. In general, notch signaling in the endocardium is an essential mediator of trabeculation by controlling proliferation and differentiation of trabecular myocytes [19]. Interestingly, when the notch ligand is endocytosed, the mechanical force applied by the bending of the membrane pulls on it and physically changes the active sight so that it is able to bind to ADAM metalloproteases. This conformational change allows γ-secretase to cleave the ligand and release the NICD [55] so it can travel to the nucleus and bind with recombination signal binding protein for the immunoglobulin k-J region (RBPJ), which is itself bound to notch target genes. While bound to notch, RBPJ acts as an activator to enhance the expression of the target gene. When notch is not present, RBPJ acts as a transcriptional repressor [19]. are two pathways: bone morphogenetic protein-10 (BMP10) in the trabecular myocardium and ephrine type-B receptor (EPHB-2)-dependent regulation of neuroregulin-1 (NRG1) in the endocardium [19, 57]. Both of these pathways are extremely important for trabeculae formation.

In brief, it has been shown that exogenous application of BMP10 can effectively rescue chick embryos that are known RBPJ knockouts [19]. These knockouts have an increase in angiogenesis in the myocardium [58] and an increase in cardiomyogenesis in embryonic stem cells when cultured in vitro [19]. The RBPJ knockout specifically in the endothelium also has a reduced expression of BMP10, endocardial EPHB2, and NRG1. The increase in cardiomyogenesis results in less trabeculation and therefore shows that notch regulates BMP10 and EPHB1-NRG1 [19].

Lee et al. solved the Navier–Stokes equations that govern the blood flow with a moving wall boundary. With their in-house, stabilized, second-order, finite element method–based flow solver, the group was able to create large-scale simulations of blood flow through zebrafish. Using their computer model, they simulated reduction of shear stress by injecting the *gata1a* MO. This resulted in a reduction of the AWSS and trabeculation [59]. To ensure that the lack of trabeculation was due to the shear force observed by the cells, the team ran the same simulation but instead changed the blood viscosity to that of the WT zebrafish. This showed that the AWSS between the WT and the adjusted *gata1a* MO zebrafish was the same and therefore supports the shear stress theory [59]. To further prove their hypothesis, the authors used *wea* mutants. This line of mutant zebrafish is unable to contract the atria, which significantly reduces the WSS across from the AV valve. Because of the lack of shear stress, trabeculation does not form. Injection of the *ErbB2* inhibitor into zebrafish only slightly reduced AWSS compared to *gata1a* MO and *wea* mutants [59]. In an attempt to restore trabeculation, both *gata1a* MO and *NICD* mRNA were injected into the zebrafish and partially restored trabeculation. However, the AWSS remained low in the dual injection fish [59].

Oscillatory fluid flow tends to be concentrated at the trabecular grooves. When pulsatile blood flow is introduced to trabeculae, the force between trabecular ridges creates a vacuum-like force that forces the blood into the cavity. The blood then travels to the bottom of the trabecular groove and up the wall of the first trabecula. Then the blood is forced down again by the pulsatory forces. Altogether, this causes oscillatory flow and therefore OSS between the trabeculae at the trabecular grooves. As the trabeculae continue to mature, the oscillatory shear force increases [59]. If *ErbB2* was inhibited, the nontrabeculated ventricle will have a lower oscillatory force, except for the region of the heart that is across from the atrium inlet [59]. Using the *gata1a* MO and the *wea* zebrafish mutants also caused a lower oscillatory force and resulting OSS at the trabecular bases. However, if *NICD* mRNA was injected as well, the oscillatory force and shear stress were similar to those of WT zebrafish [59].

8.4.3 Applications of Different Types of Shear Stress for Ventricular Morphology

Previously, our group and others have shown that WT zebrafish start trabeculation growth at 3 dpf and they continue to mature into a woven network by 5 dpf (Fig. 8.9A–D) [59]. Interestingly, by inhibiting *ErbB2*, which is downstream of notch, using AG1478 delays the initiation of trabeculation from 2 dpf in WT to 5 dpf in AG1478-injected fish (Fig. 8.9E–H) [59]. Also of interest, when mice are

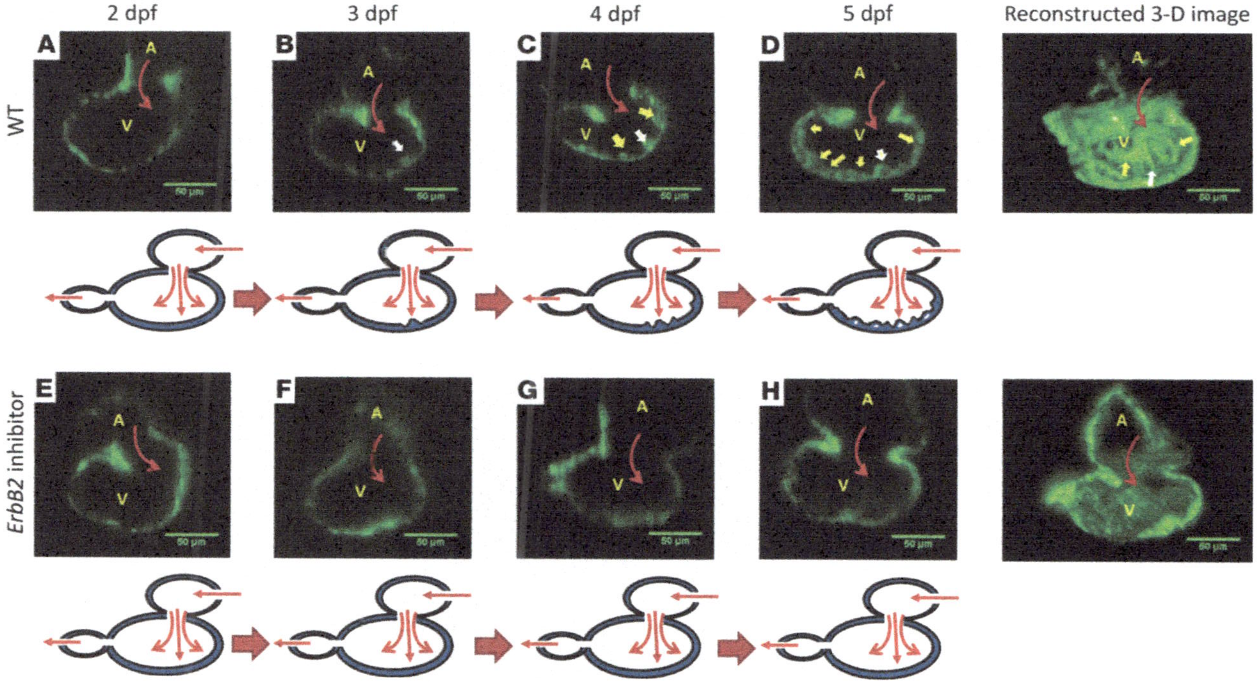

Figure 8.9 Cardiac morphogenesis of WT and ErBb2-inhibited zebrafish. (A–D) Trabeculation development through 5 dpf in WT fish. (E–H) ErbB2 inhibition caused trabeculation to be delayed through 5 dpf. Atrial blood flow (red arrows), initial trabecular ridge (white arrow), and additional trabecular ridges (yellow arrows). Reproduced from Ref. [59]. Copyright (2012) with permission from the American Society for Clinical Investigation.

lacking NRG1, they die before birth due to the lack of trabeculation and cardiac maturation [8]. Second, mice without either ERBB2 or ERBB4 also die due to defective trabeculation [8].

8.4.4 Notch Signaling for Trabeculation

In chicks, notch can be forcibly expressed using retroviruses. In particular, one group overexpressed the NICD. Interestingly, the chicks had reduced expression of the cardiac myocyte markers [19]. They also showed that overexpression of NICD in endothelial cells induces the Snail family of transcription factors. These then lead to downregulation of vascular-endothelial (VE)-cadherin and cause morphological changes similar to endothelial-mesenchymal transition (EMT) [19].

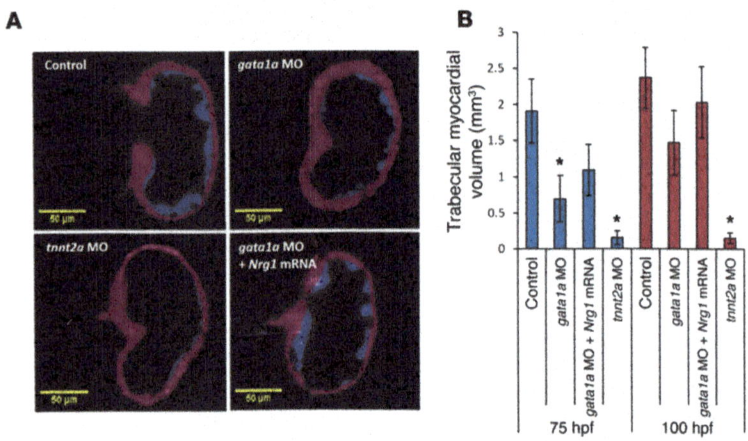

Figure 8.10 Myocardial wall thickness. (A) Myocardium visualized in Tg(cmlc:gfp) transgenic zebrafish shows outer curvature (magenta) and trabecular ridges (purple) of experimentally treated zebrafish. (B) Quantification of trabeculae volume showing that co-injection of *gata1a* MO and *Nrg1* mRNA rescued myocardial volume at 100 hpf, while *tnnt2a* MO was still significantly lower. Reproduced from Ref. [5]. Copyright (2016) with permission from the American Society for Clinical Investigation.

A second group used computational modeling from SPIM images in order to determine the effect of the forces of the size of the ventricle in zebrafish. At 75 hpf, those injected with *gata1a* MO were 2.7 ± 0.4 times smaller than the WT control. Similarly, the *tnnt2a*

MO group had a 19.0 ± 0.3 times smaller ventricle. When co-injected with *Nrg1* mRNA, the *gata1a* MO group and *tnnt2a* group were only 1.7 ± 0.4 times smaller than the control, where the *tnnt2a* MO + *Nrg1* mRNA zebrafish were similar to the *tnnt2a* MO-only group. At 100 hpf, the ventricles of *gata1a* MO, *tnnt2a* MO, and *gata1a* MO + *Nrg1* mRNA were 1.5 ± 0.4, 23.0 ± 0.3, and 1.0 ± 0.5 times smaller than the WT, respectively [5]. Interestingly, the *Nrg1* mRNA was unable to rescue the *tnnt2a* MO fish (Fig. 8.10B) [5].

8.4.5 Link Hemodynamic Shear Stress and Trabeculation: Pattern of Trabeculae

In the heart, blood flows directly onto the vessel wall opposite of the heart valve. Transverse flow has been shown to be correlated with atherosclerotic plaque formation in animals by causing endothelial cells to become misaligned [56]. Also, in areas of high stress there is continuous negative remodeling [56]. The heart compensates by increasing the surface area in the form of trabeculae to reduce the stress per area, which could lead to less detrimental effects not seen in the vessels. However, it has been shown that if the blood flow is altered, cardiac abnormalities begin to develop [60].

The primary goal of trabeculation is to make more uniform transmural stress distribution and to increase intramyocardial blood flow [61]. Additionally, myocardial trabeculae have been associated with enhancing contractility [60], ventricular septation [60], and intraventricular conduction [8] and helping to direct blood flow before septation [61]. One group in particular [60] has looked into how changing the biomechanical load on the developing heart effects normal cardiac development. The hemodynamic load was reduced by blocking all of the blood cells from entering circulation, which resulted in the heart volume and myocardial thickness becoming significantly reduced [60]. Acrylamide or TEMED was also into the blood vessels to observe the difference between normal hematocrit (TEMED) and slightly reduced hematocrit (acrylamide) on the cardiac development. It was shown that the decrease in hematocrit impairs trabeculation and cardiac looping [60]. Furthermore, circumstances where more oxygen is needed in the cardiac tissue, such as in surgical manipulations or arterial blockages, cause outward or inward remodeling that attempt to return the shear stress back to a healthy physiological range [56].

After observation that trabeculation initiates across the ventricle from a valve [6, 8], Hove et al. physically blocked the inlet or outlet to the heart of zebrafish [6]. This caused severe developmental deformations, including the absence of a valve and bulbus [6]. Additionally, cardiac looping did not occur. Blocking both the inlet and the outlet to the heart caused similar phenotypes, which brought about the conclusion that it is the lack of shear forces in the heart rather than the pressure gradient within the heart that causes these malformations [6].

Another group, Lee et al., injected *gata1a* MO into zebrafish at the 1–4 cell stage [5]. This MO reduced hematopoiesis by about 90%, which caused a reduction in the shear forces sensed by the cells. After 75 and 100 hpf, there was no trabeculation compared to the WT control [5]. Interestingly, the *gata1a* MO significantly down-regulated notch ligands DLL4, Jag1, and Jag2, receptor Notch1b, and downstream molecules Nrg1 and ErbB2 [5]. This shows that the observed reduction of trabeculation, caused by lowering shear stress, had a direct effect on the cell signaling and specifically in notch-related genes [5]. Using *Tg(flk:mCherry, tp1:gfp)* zebrafish to visualize the notch signaling specifically in the heart, they showed that the *gata1a* MO significantly reduced the shear stress in the endocardium [5]. Similarly, they used the *wea* zebrafish mutant, where atrial contraction cannot occur. The absence of atrial contraction does not produce the shear stress that usually occurs across from the AV canal. In these zebrafish, the ventricle was small and there was also no trabeculation. Additionally, the notch signaling ligands, receptor, and downstream molecules were all downregulated [5].

In hopes of restoring trabeculation, Lee et al. injected *Nrg1* mRNA into the zebrafish at the embryonic stage. In both the *wea* mutant and the *gata1a* MO zebrafish, the *Nrg1* mRNA upregulated the notch signaling–related genes at 75 and 100 hpf [5]. In addition to the upregulated notch genes, the *wea* mutant also had enhanced cardiac contraction and ventricular function [5].

The *tnnt2a* MO was also used to stop myocardial contraction. The lack of contraction would lead to a lack of cardiac strain and shear stress, which was thought to prevent trabeculation. The *tnnt2a* MO led to a significant downregulation of the notch signaling molecules and genes. At 100 hpf, the ventricle of the *tnnt2a* MO zebrafish was thin and did not have trabeculation (Fig. 8.11K) [5].

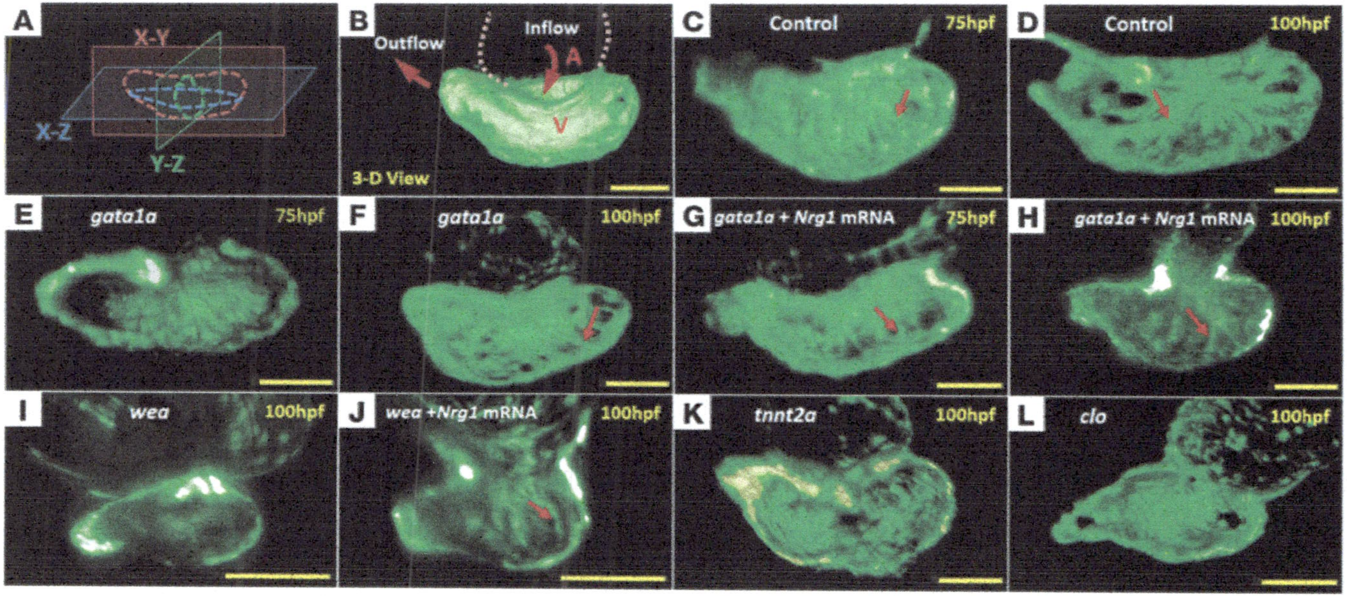

Figure 8.11 3D representation of cardiac walls in response to genetic manipulation. (A) 3D Cartesian coordinate system. (B) Inflow and outflow 3D representation. (C, D) Control 3D rendering of WT trabeculation at 75 and 100 hpf. (E, F) The *gata1a* MO–injected fish had delayed trabeculation at 75 hpf that was restored at 100 hpf. (G, H) Co-injection of *gata1a* MO and *Nrg1* mRNA restored trabeculation at 75 and 100 hpf. (I) *wea* mutant zebrafish did not have any trabeculation at 100 hpf. (J) *wea* mutants injected with *Nrg1* mRNA showed some trabeculation at 100 hpf. (K, L) *tnnta* MO and *clo* mutant showed no trabeculation at 100 hpf. Red arrows show trabecular ridges. Reproduced from Ref. [5]. Copyright (2016) with permission from the American Society for Clinical Investigation.

The *clo* zebrafish mutant develops without the endocardium, which would essentially determine whether it is the endocardium or the myocardium that senses biomechanical forces. When cultured, the *clo* mutant had significant downregulation of notch signaling, and at 100 hpf, the ventricular wall was very thin and had no trabeculation (Fig. 8.11L) [5]. Additionally, the *clo* mutant had little cardiac mRNA for notch ligands, receptor, and target genes [5]. To support the theory that it is the endothelium that senses the forces and not the lack of endothelium that causes the lack of trabeculation in the *clo* mutant, human aortic endothelial cells (HAECs) were subjected to PSS. At a force of 23 dyn cm^{-2} and a rate of 1 Hz, the PSS upregulated notch-related genes. Additionally, when notch signaling was inhibited, the notch expression returned to normal [5]. In these experiments, it is shown that the endocardium senses the shear stress caused by blood flow. Interestingly, the addition of *EPO* mRNA, which codes for blood cells, did not change the development of the trabeculae, even with an increase in viscosity. Also, the regulation of notch genes was similar to that of the control [5]. Treating the zebrafish embryos with isoproterenol to increase the contractility, and therefore the shear force, did not have a significant effect on the trabecular network or notch signaling expression [5]. The increase in heart rate could be related to athletes. Those who participate in intense cardioexercise increase their heart rate and therefore increase the shear force on the endocardium. It could be that the added shear stress is what makes exercise beneficial to heart health.

References

1. Madamanchi, N. R., Vendrov, A. and Runge, M. S. (2005). Oxidative stress and vascular disease. *Arterioscler Thromb Vasc Biol*, **25**, pp. 29–38.

2. Harrison, D., Griendling, K. K., Landmesser, U., Hornig, B. and Drexler, H. (2003). Role of oxidative stress in atherosclerosis. *Am J Cardiol*, **91**, pp. 7A–11A.

3. Hwang, J., Ing, M. H., Salazar, A., Lassegue, B., Griendling, K., Navab, M., Sevanian, A. and Hsiai, T. K. (2003). Pulsatile versus oscillatory shear stress regulates nadph oxidase subunit expression: implication for native ldl oxidation. *Circ Res*, **93**, pp. 1225–1232.

4. Li, R., Mittelstein, D., Lee, J., Fang, K., Majumdar, R., Tintut, Y., Demer, L. L. and Hsiai, T. K. (2012). A dynamic model of calcific nodule destabilization in response to monocyte- and oxidized lipid-induced matrix metalloproteinases. *Am J Physiol Cell Physiol*, **302**, pp. C658–665.

5. Lee, J., Fei, P., Sevag Packard, R. R., Kang, H., Xu, H., Baek, K. I., Jen, N., Chen, J., Yen, H., Kuo, C. C., Chi, N. C., Ho, C. M., Li, R. and Hsiai, T. K. (2016). 4-Dimensional light-sheet microscopy to elucidate shear stress modulation of cardiac trabeculation. *J Clin Invest*, **126**, p. 3158.

6. Hove, J. R., Koster, R. W., Forouhar, A. S., Acevedo-Bolton, G., Fraser, S. E. and Gharib, M. (2003). Intracardiac fluid forces are an essential epigenetic factor for embryonic cardiogenesis. *Nature*, **421**, pp. 172–177.

7. Tu, S. and Chi, N. C. (2012). Zebrafish models in cardiac development and congenital heart birth defects. *Differ Res Biol Diversity*, **84**, pp. 4–16.

8. Samsa, L. A., Yang, B. and Liu, J. (2013). Embryonic cardiac chamber maturation: trabeculation, conduction and cardiomyocyte proliferation. *Am J Med Genet Part C*, **163**, pp. 157–168.

9. Go, A. S., Mozaffarian, D., Roger, V. L., Benjamin, E. J., Berry, J. D., Borden, W. B., Bravata, D. M., Dai, S., Ford, E. S., Fox, C. S., Franco, S., Fullerton, H. J., Gillespie, C., Hailpern, S. M., Heit, J. A., Howard, V. J., Huffman, M. D., Kissela, B. M., Kittner, S. J., Lackland, D. T., Lichtman, J. H., Lisabeth, L. D., Magid, D., Marcus, G. M., Marelli, A., Matchar, D. B., McGuire, D. K., Mohler, E. R., Moy, C. S., Mussolino, M. E., Nichol, G., Paynter, N. P., Schreiner, P. J., Sorlie, P. D., Stein, J., Turan, T. N., Virani, S. S., Wong, N. D., Woo, D., Turner, M. B.; American Heart Association Statistics Committee and Stroke Statistics Subcommittee (2013). Heart disease and stroke statistics--2013 update: a report from the American Heart Association. *Circulation*, **127**, pp. e6–e245.

10. Peshkovsky, C., Totong, R. and Yelon, D. (2011). Dependence of cardiac trabeculation on neuregulin signaling and blood flow in zebrafish. *Dev Dyn*, **240**, pp. 446–456.

11. Liu, J., Bressan, M., Hassel, D., Huisken, J., Staudt, D., Kikuchi, K., Poss, K. D., Mikawa, T. and Stainier, D. Y. R. (2010). A dual role for ErbB2 signaling in cardiac trabeculation. *Development*, **137**, pp. 3867–3875.

12. Teekakirikul, P., Kelly, M. A., Rehm, H. L., Lakdawala, N. K. and Funke, B. H. (2013). Inherited cardiomyopathies: molecular genetics and clinical genetic testing in the postgenomic era. *J Mol Diagn*, **15**, pp. 158–170.

13. Moorman, A. F. M. and Christoffels, V. M. (2003). Cardiac chamber formation: development, genes, and evolution. *Physiol Rev*, **83**, pp. 158–170.
14. Weiford, B. C., Subbarao, V. D. and Mulhern, K. M. (2004). Noncompaction of the ventricular myocardium. *Circulation*, **109**, pp. 2965–2971.
15. Jenni, R., Oechslin, E. N. and van der Loo, B. (2007). Isolated ventricular non-compaction of the myocardium in adults. *Heart*, **3**, pp. 11–15.
16. Chi, N. C., Shaw, R. M., Jungblut, B., Huisken, J., Ferrer, T., Arnaout, R., Scott, I., Beis, D., Xiao, T., Baier, H., Jan, L. Y., Tristani-Firouzi, M. and Stainier, D. Y. R. (2008). Genetic and physiologic dissection of the vertebrate cardiac conduction system. *PLoS Biol*, **6**, p. e109.
17. Hu, N., Yost, H. J. and Clark, E. B. (2001). Cardiac morphology and blood pressure in the adult zebrafish. *Anat Rec*, **264**, pp. 1–12.
18. Lee, J., Cao, H., Kang, B. J., Jen, N., Yu, F., Lee, C. A., Fei, P., Park, J., Bohlool, S., Lash-Rosenberg, L., Shung, K. K. and Hsiai, T. K. (2014). Hemodynamics and ventricular function in a zebrafish model of injury and repair. *Zebrafish*, **11**, pp. 447–454.
19. High, F. A. and Epstein, J. A. (2008). The multifaceted role of notch in cardiac development and disease. *Nat Rev Genet*, **9**, pp. 49–61.
20. Yu, F., Huang, J., Adlerz, K., Jadvar, H., Hamdan, M. H., Chi, N., Chen, J. N. and Hsiai, T. K. (2010). Evolving cardiac conduction phenotypes in developing zebrafish larvae: implications to drug sensitivity. *Zebrafish*, **7**, pp. 325–331.
21. Yu, F., Li, R., Parks, E., Takabe, W. and Hsiai, T. K. (2010). Electrocardiogram signals to assess zebrafish heart regeneration: implication of long qt intervals. *Ann Biomed Eng*, **38**, pp. 2346–2357.
22. Milan, D. J., Jones, I. L., Ellinor, P. T. and MacRae, C. A. (2006). In vivo recording of adult zebrafish electrocardiogram and assessment of drug-induced qt prolongation. *Am J Physiol Heart Circul Physiol*, **291**, pp. H269–273.
23. Bakkers, J. (2011). Zebrafish as a model to study cardiac development and human cardiac disease. *Cardiovasc Res*, **91**, pp. 279–288.
24. Poss, K. D. (2007). Getting to the heart of regeneration in zebrafish. *Semin Cell Dev Biol*, **18**, pp. 36–45.
25. Poss, K. D., Wilson, L. G. and Keating, M. T. (2002). Heart regeneration in zebrafish. *Science*, **298**, pp. 2188–2190.
26. Vermot, J., Forouhar, A. S., Liebling, M., Wu, D., Plummer, D., Gharib, M. and Fraser, S. E. (2009). Reversing blood flows act through klf2a to ensure normal valvulogenesis in the developing heart. *PLoS Biol*, **7**, p. e1000246.

27. Yuste, R. (2005). Fluorescence microscopy today. *Nat Methods*, **2**, pp. 902–904.
28. Olarte, O. E., Andilla, J., Gualda, E. J. and Loza-Alvarez, P. (2018). Light-sheet microscopy: a tutorial. *Adv Opt Photonics*, **10**, pp. 111–179.
29. Chatterjee, K., Pratiwi, F. W., Wu, F. C. M., Chen, P. and Chen, B. C. (2018). Recent progress in light sheet microscopy for biological applications. *Appl Spectrosc*, **72**, pp. 1137–1169.
30. Fei, P., Lee, J., Packard, R. R., Sereti, K. I., Xu, H., Ma, J., Ding, Y., Kang, H., Chen, H., Sung, K., Kulkarni, R., Ardehali, R., Kuo, C. C., Xu, X., Ho, C. M. and Hsiai, T. K. (2016). Cardiac light-sheet fluorescent microscopy for multi-scale and rapid imaging of architecture and function. *Sci Rep*, **6**, p. 22489.
31. Dodt, H. U., Leischner, U., Schierloh, A., Jahrling, N., Mauch, C. P., Deininger, K., Deussing, J. M., Eder, M., Zieglgansberger, W. and Becker, K. (2007). Ultramicroscopy: three-dimensional visualization of neuronal networks in the whole mouse brain. *Nat Methods*, **4**, pp. 331–336.
32. Greger, K., Swoger, J. and Stelzer, E. H. (2006). Basic building units and properties of a fluorescence single plane illumination microscope. *Rev Sci Instrum*, **78**, p. 023705.
33. Huisken, J. and Stainier, D. Y. R. (2009). Selective plane illumination microscopy techniques in developmental biology. *Development*, **136**, pp. 1963–1975.
34. Ding, Y., Lee, J., Ma, J., Sung, K., Yokota, T., Singh, N., Dooraghi, M., Abiri, P., Wang, Y., Kulkarni, R. P., Nakano, A., Nguyen, T. P., Fei, P. and Hsiai, T. K. (2017). Light-sheet fluorescence imaging to localize cardiac lineage and protein distribution. *Sci Rep*, **7**, p. 42209.
35. Ding, Y., Yu, J., Abiri, A., Lee, J., Chang, C.-C., Baek, K. I., Packard, R. and Hsiai, T. (2018). Integrating 4 D light-sheet imaging with interactive virtual reality to recapitulate developmental cardiac mechanics and physiology. *Proc SPIE BiOS*, **10484**, pp. 104841C-104841-104841C-104848.
36. Peng, F., Nie, J., Lee, J., Ding, Y., Li, S., Zhang, H., Hagiwara, M., Yu, T., Segura, T., Ho, C.-M., Zhu, D. and Hsiai, T. K. (2019). Subvoxel light-sheet microscopy for high-resolution high-throughput volumetric imaging of large biomedical specimens. *Adv Photonics*, **1**, pp. 016002-016001-016002-016013.
37. Rocha, M. D., During, D. N., Bethge, P., Voigt, F. F., Hildebrand, S., Helmchen, F., Pfeifer, A., Hahnloser, R. H. R. and Gahr, M. (2019). Tissue clearing and light sheet microscopy: imaging the unsectioned adult zebra finch brain at cellular resolution. *Front Neuroanat*, **13**, p. 13.

38. Vedula, V., Lee, J., Xu, H., Kuo, C. J., Hsiai, T. K. and Marsden, A. L. (2017). A method to quantify mechanobiologic forces during zebrafish cardiac development using 4-D light sheet imaging and computational modeling. *PLoS Comput Biol*, **13**, p. e1005828.

39. Fuchs, E., Jaffe, J. S., Long, R. A. and Azam, F. (2002). Thin laser light sheet microscope for microbial oceanography. *Opt Express*, **10**, pp. 145–154.

40. Keller, P. J. and Dodt, H. U. (2012). Light sheet microscopy of living or cleared specimens. *Curr Opin Neurobiol*, **22**, pp. 138–143.

41. Glaser, A. K., Reder, N. P., Chen, Y., McCarty, E. F., Yin, C., Wei, L., Wang, Y., True, L. D. and Liu, J. T. C. (2017). Light-sheet microscopy for slide-free non-destructive pathology of large clinical specimens. *Nat Biomed Eng*, **1**, p. pii:0084.

42. Dworetsky, M. (1983). A period-finding method for sparse randomly spaced observations or "how long is a piece of string?". *Mon Not R Astron Soc*, **203**, pp. 917–924.

43. Stellingwerf, R. (1978). Period determination using phase dispersion minimization. *Astrophys J*, **224**, pp. 953–960.

44. Thevenaz, P., Ruttimann, U. E. and Unser, M. (1998). A pyramid approach to subpixel registration based on intensity. *IEEE Trans Image Process*, **7**, pp. 27–41.

45. Liebling, M., Forouhar, A. S., Gharib, M., Fraser, S. E. and Dickinson, M. E. (2005). Four-dimensional cardiac imaging in living embryos via postacquisition synchronization of nongated slice sequences. *J Biomed Opt*, **10**, p. 054001.

46. Marsden, A. L., Feinstein, J. A. and Taylor, C. A. (2008). A computational framework for derivative-free optimization of cardiovascular geometries. *Comput Methods Appl Mech Eng*, **197**, pp. 1890–1905.

47. Moghadam, M. E., Bazilevs, Y., Hsia, T.-Y., Vignon-Clementel, I. E. and Marsden, A. L. (2011). A comparison of outlet boundary treatments for prevention of backflow divergence with relevance to blood flow simulations. *Comput Mech*, **48**, pp. 277–291.

48. Myronenko, A. and Song, X. (2010). Intensity-based image registration by minimizing residual complexity. *IEEE Trans Med Imaging*, **29**, pp. 1882–1891.

49. Esmaily-Moghadam, M., Bazilevs, Y. and Marsden, A. L. (2015). A bi-partitioned iterative algorithm for solving linear systems arising from incompressible flow problems. *Comput Methods Appl Mech Eng*, **286**, pp. 40–62.

50. Arbia, G., Corsini, C., Moghadam, M. E., Marsden, A. L., Migliavacca, F., Pennati, G., Hsia, T.-Y., Vignon-Clementel, I. E.; Investigators MOCHA (2014). Numerical blood flow simulation in surgical corrections: what do we need for an accurate analysis? *J Surg Res*, **186**, pp. 44–55.

51. Steinman, D. A., Hoi, Y., Fahy, P., Morris, L., Walsh, M. T., Aristokleous, N., Anayiotos, A. S., Papaharilaou, Y., Arzani, A. and Shadden, S. C. (2013). Variability of computational fluid dynamics solutions for pressure and flow in a giant aneurysm: the ASME 2012 summer bioengineering conference CFD challenge. *J Biomech Eng*, **135**, p. 021016.

52. Lee, J., Moghadam, M. E., Kung, E., Cao, H., Beebe, T., Miller, Y., Roman, B. L., Lien, C. L., Chi, N. C., Marsden, A. L. and Hsiai, T. K. (2013). Moving domain computational fluid dynamics to interface with an embryonic model of cardiac morphogenesis. *PLoS One*, **8**, p. e72924.

53. Esmaily-Moghadam, M., Hsia, T.-Y., Marsden, A. L. and Modeling of Congenital Hearts Alliance I. (2015). The assisted bidirectional Glenn: a novel surgical approach for first-stage single-ventricle heart palliation. *J Thorac Cardiovasc Surg*, **149**, pp. 699–705.

54. Löhner, R. (1995). Robust, vectorized search algorithms for interpolation on unstructured grids. *J Comput Phys*, **118**, pp. 380–387.

55. D'Amato, G., Luxán, G. and de la Pompa, J. L. (2016). Notch signalling in ventricular chamber development and cardiomyopathy. *FEBS J*, **283**, pp. 4223–4237.

56. Baeyens, N., Bandyopadhyay, C., Coon, B. G., Yun, S. and Schwartz, M. A. (2016). Endothelial fluid shear stress sensing in vascular health and disease. *J Clin Invest*, **126**, pp. 821–828.

57. Teichmann, U. and Kessel, M. (2004). Highly restricted bmp10 expression in the trabeculating myocardium of the chick embryo. *Dev Genes Evol*, **214**, pp. 96–98.

58. Diaz-Trelles, R., Scimia, M. C., Bushway, P., Tran, D., Monosov, A., Monosov, E., Peterson, K., Rentschler, S., Cabrales, P., Ruiz-Lozano, P. and Mercola, M. (2016). Notch-independent rbpj controls angiogenesis in the adult heart. *Nat Commun*, **7**, p. 12088.

59. Lee, J., Vedula, V., Baek, K. I., Chen, J., Hsu, J. J., Ding, Y., Chang, C.-C., Kang, H., Small, A., Fei, P., Chuong, C.-M., Li, R., Demer, L., Packard, R. R. S., Marsden, A. L. and Hsiai, T. K. (2018). Spatial and temporal variations in hemodynamic forces initiate cardiac trabeculation. *JCI Insight*, **3**, p. pii:96672.

60. Hoog, T. G., Fredrickson, S. J., Hsu, C. W., Senger, S. M., Dickinson, M. E. and Udan, R. S. (2018). The effects of reduced hemodynamic loading

on morphogenesis of the mouse embryonic heart. *Dev Biol*, **442**(1), pp. 127–137.

61. Bartman, T. and Hove, J. (2005). Mechanics and function in heart morphogenesis. *Dev Dyn*, **233**, pp. 373–381.
62. Li, R., et al. (2015). Disturbed flow induces autophagy, but impairs autophagic flux to perturb mitochondrial homeostasis. *Antiox Redox Signal*, **23**(15), pp. 1207–1219.

Index